SOUTHEAST ASIAN ENERGY TRANSITIONS

In memory of Nellie
For Paul

Southeast Asian Energy Transitions

Between Modernity and Sustainability

MATTIJS SMITS
Wageningen University, The Netherlands

LONDON AND NEW YORK

First published 2015 by Ashgate Publishing

2 Park Square, Milton Park, Abingdon, Oxfordshire OX14 4RN
711 Third Avenue, New York, NY 10017

Routledge is an imprint of the Taylor & Francis Group, an informa business

First issued in paperback 2018

British Library Cataloguing in Publication Data
A catalogue record for this book is available from the British Library.

The Library of Congress has cataloged the printed edition as follows:
Smits, Mattijs.
  Southeast Asian energy transitions : between modernity and sustainability / by Mattijs Smits.
    pages cm
  Includes bibliographical references and index.
  ISBN 978-1-4724-4875-0 (hardback) 1. Energy industries–Southeast Asia. 2. Renewable energy sources–Southeast Asia. 3. Carbon dioxide mitigation–Southeast Asia. I. Title.
  HD9502.A782S65 2015
  333.790959–dc23

                                                                                                2015005950

ISBN 978-1-4724-4875-0 (hbk)
ISBN 978-1-138-54685-1 (pbk)

Contents

List of Figures

List of Tables

Notes on the Author

Mattijs Smits researches and teaches in the fields of energy policy and politics, environment, sustainability, (rural) development and climate finance. During his academic and professional career, he spent extended periods living and working as a researcher and consultant in Southeast Asia, notably in Laos, Thailand and Vietnam. He holds degrees from four different universities on three continents: a BSc from the University of Utrecht, an MSc from Wageningen University and PhD degrees from The University of Sydney and Chiang Mai University (as part of a cotutelle arrangement). He is currently working as an assistant professor at the Environmental Policy Group of Wageningen University, The Netherlands.

Foreword

Energy options and practices are integral to the process of development. Energy is also at the centre of concerns over climate change, various dimensions of geopolitical and livelihood security, global-level land-use choices, and other present day debates on global futures.

Perhaps nowhere are questions over energy futures more pertinent than in the rapidly growing economies of Southeast Asia, and in particular in the countries of the Mekong Region. Not only does rapid economic development produce pressures for individual countries to exploit their own energy resources, with accompanying social and environmental impacts. Energy development is also integral to the relationship between countries, as electricity is sold across borders and as energy development in some states becomes not just a means to power economic activity but a source of national income in its own right. And energy is integral to the changes in daily life that define the experience of development at a micro-level.

There is an understandable tendency to focus on the material economic and environmental questions around power production. Peak oil, impacts of dams, winners and losers in the energy game are all much discussed and relevant issues. Yet the understanding of energy questions in these terms tends to miss the deeper ways in which energy is imbued within bigger questions of development.

In this book, Mattijs Smits gets to the very heart of two big themes in development: modernity and sustainability. These themes are all too often understood in contraposition to one another. In the long-run narrative of trends in development studies, modernity gets associated with the early faith in large-scale development associated with modernisation theory of the early development era, while sustainability is posited as a late-development era reaction to, or taming of, the growth-at-all-costs approach. As Smits shows eloquently in this book, however, modernity and sustainability are much better understood as complementary and even mutually constitutive concepts. He illustrates this mutuality clearly in the context of energy choices and debates.

What distinguishes this book, however, is not just the big ideas that it engages. It is also marked by two significant departures in work on engagement of energy issues with development studies.

The first key departure of the book is the framing of the work in social theory. Smits employs the concept of energy transitions to apply ideas around socio-technical studies, ecological modernisation and political ecology to the context of change, choices and the tensions in each. He also sets a convincing agenda for a multi-scalar and nuanced approach to geographies of energy – and more specifically electricity – development.

The second key departure of Mattijs Smits's fine contribution is its grounding in country and village level case studies. Employing two cases from Thailand and two from Laos, each based on detailed fieldwork, Smits shows how countries that share commonalities of history and culture, share a common border, and that are intricately linked in energy and associated economic interdependence, nevertheless manifest framings and debates around energy development that are associated with quite specific trajectories of social, economic and political development. At a local level, the embedding of energy transitions in everyday life is a particularly significant aspect of the ethnographic component to Smits's work.

This book is far from the last word in writing on energy transitions and their association with the tensions between modernity and sustainability. On the contrary, the approach lends itself to thinking about such issues well beyond Thailand and Laos, the Mekong, and Southeast Asia, and as such has relevance to all those interested in the embeddedness of energy within societal transitions associated with development.

Philip Hirsch
The University of Sydney, Australia

Preface

This book addresses two important sets of questions. First, how can we reconcile development and growth of energy demand with questions about sustainability? Can we maintain and increase economic prosperity – driven by increased energy usage – while lowering our environmental impact and carbon footprint? Is it possible to eradicate poverty without disturbing the planet's ecosystems even further? These are questions that have been on the agenda for decades, but are still highly relevant in times of persistent poverty, climate change and crises. Second, what roles do energy and sustainability play in driving social and political change? How does access and availability of energy (continue to) change lives in the global North and South? What is the interaction between energy, technology and the way we travel, communicate and entertain ourselves? This second group of questions has been surprisingly underexposed, but is – in my view – critical in order to find answers to the first set.

In order to tackle these questions, this book switches gears between, on the one hand, academic debates centred around modernity and (energy) transition, and, on the other hand, detailed empirical investigations of these issues at regional, national and local scales in Southeast Asia. In this region, we find a wide variety of countries, peoples and livelihoods; from devastatingly poor to incredibly rich, from small upland ethnic minorities to rich cosmopolitan urbanites, and everything in between. In terms of energy, there is a similar diversit; from large-scale coal and hydropower plants to household-scale solar and hydropower systems, and from power-guzzling shopping malls to people living without access to electricity or other 'modern' forms of energy. Most Southeast Asian examples in this book come from Thailand and Laos, two neighbouring countries harbouring these enormous varieties – while sharing borders, languages and cultures – making it a very rich context for research. The concepts, issues and approaches used, however, are not unique to Southeast Asia and are therefore relevant to other areas of the world.

On a personal level, this book represents a milestone in my work on energy and development issues in Southeast Asia. Since 2007, I have lived and worked for extended periods of time in Laos and Thailand, while travelling to many other Southeast Asian countries. During this time, I researched diverse sets of issues, such as small-scale renewable energy systems, the politics of energy policy, and sustainability transitions. I developed a grounded fieldwork-based approach, while engaging with global literature and debates. This book consolidates these insights and should therefore appeal to a broad audience – academics, students, policy makers and lay audiences – interested in sustainability transitions, (renewable) energy technology and policy, development, and interdisciplinary social science, and social theory.

Mattijs Smits

Acknowledgements

This book would not be possible without the help and support of a large number of people all over the world. The majority of this book's ideas took shape at The University of Sydney with brilliant academic support from Philip Hirsch, Phil McManus and Tira Foran (CSIRO), and through many interactions with my friends and colleagues there. Benjamin Sovacool encouraged me to publish this work, and the people at Ashgate, especially Katy Crossan, provided great help to realise this book.

There are too many people to thank in Southeast Asia, notably the countless people who so generously made time to sit through long interviews, walked me through their villages, and provided food and a place to sleep during fieldwork. Busarin (Aer), Lada, Nith, Somsy, Xoryang and Yeeyang provided essential help during data collection. Part of the work would not be possible without the staff and students of Chiang Mai University, in particular Ajarn Yos Santasombat and Ajarn Santita Ganjanapan. In addition, there are those who helped with ideas, logistics, feedback or all of the above: Carl Middleton, Charoensri Huadmai, Chris and Chom Greacen, Daniel Robinson, Gavin Bridge, Hanna Kaisti, Jakob Rietzler, Jiab Tongsopit, Leon Gaillard, Louis Lebel, Mira Käkönen, Nishan Disanayake, Peter du Pont, Sabrina Gyorvary, Samuel Martin, Witoon Permpongsacharoen and his MEE Net team, the management and staff of Sunlabob and LIRE, and many others.

I should also thank my colleagues of the Environmental Policy Group at Wageningen University, in particular Simon Bush, who can take credit for introducing me to Southeast Asia and to publishing in general. I also thank Arthur Mol for supporting my return to Wageningen and this book project.

Finally, a big thank you to my family and friends, starting with my brothers and sisters – Peter, Marije, Klara, Tomas and Judith – and their partners. Laurie, my stepmother, thank you for your unconditional support and care. To Michelle: 'less than three'! Finally, I am proud to dedicate this book to my late mother, Nellie, and my father Paul.

List of Abbreviations

ADB	Asian Development Bank
ANT	Actor-Network Theory
ASEAN	Association of Southeast Asian Nations
CFL	Compact Fluorescent Lamp
DEDE	Department of Alternative Energy Development and Efficiency (Thailand)
DSM	Demand-Side Management
EdL	Electricité du Lao
EGAT	Electricity Generating Authority of Thailand
EMT	Ecological Modernisation Theory
EPPO	Energy Policy and Planning Office (Thailand)
ERC	Energy Regulatory Commission (Thailand)
GDP	Gross Domestic Product
GMS	Greater Mekong Subregion
GWh	Gigawatt-hour (10^9 Watt expended over 1 hour)
IEA	International Energy Agency
IMF	International Monetary Fund
IPP	Independent Power Producer
kW	Kilowatt (10^3 Watt)
kWh	Kilowatt-hour (10^3 Watt expended over 1 hour)
LAK	Lao Kip (Lao currency)
Lao PDR	Lao People's Democratic Republic
LPRP	Lao Peoples' Revolutionary Party
LTS	Large Technical Systems
MEA	Metropolitan Electricity Authority (Thailand)
MEM	Ministry of Energy and Mines (Laos)
MLP	Multi-Level Perspective
MW	Megawatt (10^6 Watt)
NEA	National Energy Authority (Thailand)
NEPO	National Energy Policy Office (Thailand)
NGO	Non-governmental organisation
NTFP	Non-timber forest product
PDEM	Provincial Department of Energy and Mines (Laos)
PDP	Power Development Plan
PEA	Provincial Electricity Authority (Thailand)

PPA	Power Purchase Agreement
PV	Photovoltaic
RISE	Rural Income through Sustainable Energy
SAO	Sub-district (tambon) Administrative Organisation (Thailand)
SCOT	Social Construction of Technology
SHS	Solar Home Systems
SPP	Small Power Producer
STS	Science and Technology Studies
THB	Thai Baht (Thai currency)
TIS	Technological Innovation Systems
UNDP	United Nations Development Programme
USAID	United States Agency for International Development
VSPP	Very Small Power Producers
W	Watt

Chapter 1
Energy, Modernity and Sustainability in Southeast Asia

Setting the Scene

Figure 1.1   **Picture from the Electricité du Laos (EdL) Annual Report 2010. Original caption: 'Construction of 230 kV transmission line Nam Theun Hinboun Expansion Project'**

Source: EdL (2011b), reproduced with permission.

This book addresses the apparent tensions between modernity and sustainability in energy transitions through the study of power sector developments in Southeast Asia at different scales. A good starting point is Figure 1.1, which pictures a woman in Laos – one of the 'least developed countries' in Southeast Asia – carrying a load of firewood, presumably for cooking and possibly to warm her house. It also shows transmission lines from an expansion of the first completed Independent Power Producer (IPP) project in Laos, the Theun-

Hinboun hydropower project. This project exports more than 99 per cent[1] of the generated electricity to neighbouring Thailand, an upper middle-income country and the second largest economy in Southeast Asia. Nevertheless, 21 per cent of households and 31 per cent of all villages in Laos lacked access to electricity in 2011 (EdL, 2012b). In contrast, Thailand has not built any hydropower plants since 1994, not only due to lack of sites, but also because of the increasing domestic opposition against large-scale power plants. These two elements provide a snapshot of the issues and contrasts in the energy sector in Southeast Asia. One objective of this book is to provide a scalar analysis of how such historical mutual energy dependences developed and what the consequences have been and continue to be for peoples' livelihoods.

Another striking aspect of Figure 1.1 is that it featured in an annual report of Electricité du Laos (EdL, 2011b), the state-owned electricity utility in Laos.[2] Thus, EdL's interpretation of the picture seems to be to show the rapid development of the power sector in Laos and the modernisation of the country, rather than to imply any inequity between modern power lines and the woman in the foreground. Pictures like these – depicting electricity infrastructure such as dams and transmission lines – also figure on the website of the utility, on its calendar and in other reports. The link between energy and modernity constitutes the other important backbone of this book.

Between Modernity and Sustainability

This study investigates the tensions between modernity and sustainability through the lens of developments in the energy sectors of Southeast Asia,[3] with a specific focus on Thailand and Laos. Since the oil crises of the 1970s, the link between energy and the world economy has become increasingly clear in media and public policy discourses. Around the same time, awareness of the local and global environmental and social impacts of energy production and consumption and the need for 'sustainable development' started to emerge (Brundtland, 1987; Meadows et al., 1974). In particular, over the last decade, focus has been on climate change. Notwithstanding, the global primary energy

---

1   Average amount of electricity exported between 1999 and 2011 is 99.3 per cent (EdL, 2012b).

2   Since 2010, EdL-Generation has operated as a subsidiary to the state-owned company EdL. In 2011, EdL-Generation was one of the first two companies on the newly-opened Lao Stock Exchange. 10 per cent of its shares were sold to the Thai-based company Ratchaburi Electricity PCL, 15 per cent sold domestically and 75 per cent remained with EdL (Asia Times Online (Clifford McCoy), 2011; Hookway, 2011).

3   The 11 Southeast Asian countries are: Brunei Darussalam, Cambodia, Indonesia, Lao PDR, Malaysia, Myanmar, Philippines, Singapore, Thailand, Timor-Leste and Vietnam.

demand has increased on average by more than 2 per cent per year since the 1970s and is projected to continue to grow (IEA, 2000, 2012). Despite this increased demand, 1.3 billion people are still without access to electricity and 2.6 billion without access to 'clean cooking facilities'. These absolute numbers are expected to reduce only slightly by 2030, according to the World Energy Outlook for 2012 (IEA, 2012).

This research investigates some of the seeming paradoxes underpinning these statistics in Southeast Asia. Thailand and Laos represent both the diversity and the similarities in terms of political system and development within this subcontinent. Over time, colonialism and the wars in Indochina separated these two countries politically. Landlocked Laos, on the one hand, has had a socialist government since 1975 and belonged to the UN list of four Least Developed Countries in Southeast Asia in 2014, along with Cambodia, Myanmar and Timor-Leste. Thailand, on the other hand, has had alternating democratic and military governments since 1932, but had the third highest GDP per capita in the region, after Singapore and Malaysia. However, the majority of the people in Thailand and Laos share a common culture and religion, a similar language and shared market-based economies, since the opening of the Lao economy following the collapse of the Soviet Union. In terms of energy, the two countries have become increasingly mutually dependent, with Thai companies and banks financing and constructing hydropower and other energy projects in Laos. The electric power generated and exported to the Thai electricity market provides the foreign exchange, which is deemed very important for a landlocked country like Laos with limited export opportunities. Moreover, the demand for electricity is increasing rapidly in both countries, outpacing economic growth and strengthening the discourse of regional trade and cooperation which has emerged over the last few decades in Southeast Asia. On the back of these developments, increasingly mobility and communication challenge the existing culture and create new and hybrid cultures.

Energy Transitions, Modernity and Sustainability

Against this background, the aim of this book is to unpack the tensions between modernity and sustainability by looking at energy sector developments in Southeast Asia on different scales: regional, national and local. The book mainly focuses on electricity as an energy carrier; but it also provides some coverage of the changes in fuels for cooking and mobility. The approach of this book is geographic and multi-disciplinary and it draws upon disciplines such as human geography, (environmental) sociology, political ecology, anthropology, development studies, science and technology studies, and social theory more generally.

The use and choice of 'modernity' as one of the key concepts of this book is discussed in detail in Chapter 2, but I will briefly introduce it here. Modernity, in effect, can be distinguished as two distinct ideal-types: (1) as a philosophical or epistemological condition; and (2) as a distinct historical or empirical instance (Wagner, 2001). Both conceptualisations are important for this book. The former meaning will be used when referred to in the analysis of literature and social theory as a lens through which to explore the changing human condition, or, more precisely, the processes of development and social change in energy sector developments in Southeast Asia. This type of conceptualising modernity has deep roots in the social sciences, for example, in the philosophies of Durkheim and Marx, and it is strongly related to the *zeitgeist* (Dodd, 1999). However, the role of energy, or other environmental and material conditions, is often underplayed when discussing this type of modernity.

The use of modernity as historical or empirical instance is important insofar as this term – or its derivatives modern and modernisation – is used in everyday vocabulary to distinguish it from traditional or 'less modern' ways of doing things. This last conceptualisation of modernity recognises the fact that modernity has different meanings for different actors. Thus, it may be used to analyse discourses about tradition and change related to livelihood change and developments in the energy sector. These discourses of modernity, however, are not just 'language games', but they have important material consequences, such as in the organisation of infrastructure, homes, power stations, electricity grids and electrical appliances (Graham and Marvin, 1994; Hughes, 1983). Finally, it is important to recognise that these two conceptualisations of modernity are ideal-types; in practice and academic literature, they might overlap and take hybrid forms.

This book argues that energy and both conceptualisations of modernity are closely intertwined. Indeed, society cannot be understood without showing the importance of changes in energy production, conversion, transmission, distribution and consumption for human development and civilisation (Smil, 1994). Moreover, as has been shown in Figure 1.1, 'modern' energy uses, and electricity in particular, have a strong role in advocating a specific form of modernity. As Bridge (2011) contends:

[e]nergy is one of modernity's fundamental mediums and metrics. Defined as the capacity to do work, energy is the productive force at the heart of many economic, social and environmental changes associated with modernist transformation … Governments seeking short cuts to modernity plough resources into nuclear power, large-scale hydro, and rural electrification as a means of spreading the light of development across a territory, drawing citizens out of the metaphorical darkness of tradition, and forging a national imaginary. (p. 307)

4

In line with the above statement, this book argues that modernity and energy development are intimately related, not only in terms of economic developments, but also in terms of human-environment interactions and changing livelihoods and practices. Rather than taking modernity at face value, this research investigates its different guises, both in the academic literature as well as its various manifestations in discourses by actors. Moreover, it aims to explore the tensions between energy, modernity and sustainability and its various manifestations in the context of energy transitions in Southeast Asia.

This book also engages with sustainability, another widely used and contested concept. Much like modernity, the concept of sustainability carries different meanings in different circumstances. In this book, it is understood as environmental sustainability, which can be defined as 'the degree to which a process or enterprise is able to be maintained or continued while avoiding the long-term depletion of natural resources' (Oxford English Dictionary, 2012b). While there is probably little disagreement over such a general definition, specifying what this means for specific processes, in specific contexts, and in specific timeframes reveals the contested nature of this concept. As such, sustainability has been framed as one of the most important challenges in the last few decades as well as 'a major intellectual and research agenda for human geography' (Marsden, 2009, p. 11: 104).

The key challenge related to environmental sustainability in this book is to explore its real and perceived tensions with modernity in the context of energy sector developments in Southeast Asia. Simply put, on the one hand, sustainability appears distinctly opposite to energy-modernity, because history has shown that primary energy use has steadily risen for many centuries now, resulting in increasingly pressing problems of pollution, environmental degradation and climate change. On the other hand, people have argued that problems of un-sustainability can be 'solved' with increased modernisation. This notion, which is at the core of the concept of 'sustainable development' (Brundtland, 1987), was a deliberate attempt to resolve the perceived tensions between development and sustainability. While highly influential, it also received widespread criticism (see Redclift, 2005, for a comprehensive overview) for its inability to solve environmental problems and perpetuate existing power relations. That is why this book frames energy-related challenges as a problem of sustainability and modernity, rather than of 'development'. The next section shows how these ideas have been translated into a context-specific research aim and questions.

The Aim of the Book

The main aim of this book is to explore the tensions, relations and interactions between modernity and sustainability through analysis of energy transitions and trajectories in the energy sectors in Southeast Asia. Specifically, this book explores these tensions and interactions through case studies at different scales: regional level (ASEAN and Greater Mekong Subregion), national level (Thailand and Laos) and local level (three villages and one sub-district). This has given rise to the following three questions:

i) How have state formation, globalisation, territorialisation and regional integration shaped – and been shaped by – energy transitions in Southeast Asia, Thailand and Laos since the mid-nineteenth century?
ii) How are changing energy production and consumption embedded in socio-economic development and changing livelihoods in local energy trajectories in case studies in Thailand and Laos?
iii) How do these energy trajectories influence energy practices, cultures and worldviews?

A Social Scientific Framing of Energy Transitions

Epistemological Framing

This research draws upon a wide range of literature and perspectives of the social sciences to tackle the broad questions of modernity and sustainability. As such, it is more issue-oriented than discipline-based. In doing so, it is broadly sympathetic to the idea of inter-disciplinary, if not post-disciplinary, research (Sayer, 2000a). It also favours flexible methodologies which can be adapted to the highly interconnected and interdependent human-environment challenges the world is facing today (Law, 2004).

There are two metaphysical positions that underpin the epistemological framing of this book: critical realism and actor-network theory.[4] While this research does not attempt to resolve the ontological differences between these two positions, there are a number of similar epistemological consequences. Besides their commitment to cut across different disciplines, both aim to find the middle ground between strong positivism and relativism. Critical realism does so by employing a relativist epistemology and moderate non-essentialism, while at the same time adhering to a realist ontology (Sayer, 1997, 2000b). Actor-network theory, on the other hand, proposes a flat ontology whereby humans and their social constructions are not privileged in any way (Law, 1992). Both

---

4 Some of the key authors associated with actor-network theory have criticised this name explicitly (Latour, 1999) and proposed 'material semiotics' as a better alternative (Law, 2009). However, this name is not used much in other literature.

positions leave space for the analysis of subjectivity and discourse, without lapsing into the conclusion that everything is socially constructed.

A second and related similarity is the focus on the role of objects and materiality. Actor-network theory in particular emphasises the symmetry of human and non-human actors in order to understand the formation of assemblages made up of material and expressive elements (DeLanda, 2006) or quasi-objects (Latour, 1993). The realism in critical realism informs this metaphysics about the role of material objects in shaping and being shaped by their human interactions (Forsyth, 2001). These ideas are important for the analysis of energy systems, which involve large-scale infrastructural networks and high levels of path dependency (Graham and Marvin, 1994).

A third similarity is their conceptualisation of causality, which goes beyond the 'billiard-ball' conceptualisation wherein certain events always and linearly result in fixed outcomes. Instead, for critical realists, a certain outcome is the effect of a mechanism which operates in a certain context (Sayer, 2000b). Because the context is – in the social sciences – of necessity varying or unknown, the mechanism might produce different outcomes or not work altogether. Actor-network theory is committed to the idea of emergence, which is the result of a dispersed agency and cannot be attributed to specific actors, but rather to the relations and networks they form (Ingold, 2008). These models of causality help to question complex relations such as those between energy demand and economic growth or the ways in which modernity plays a role in the type of form of energy infrastructure.

Finally, both critical realism and actor-network theory are empirically, rather than theoretically, oriented and open to diverse and flexible methodologies. Sayer (2000b), for example, shows that critical realism in the social sciences is open to the use of both quantitative and qualitative methods and triangulation, the use of which depends on the research problem. For actor-network theorists, the world consists of hybrid networks of different objects, people and quasi-objects. This means that research methodologies in general should try to follow these actors (Latour, 1987), rather than trying to come up with methodologies based on artificial boundaries between nature and society, quantitative and qualitative, and micro and macro. While it is difficult to avoid using these categories, this book nonetheless uses a variety of different methods in different situations and triangulates wherever possible (Law, 2004).

Four Social Science Perspectives on Energy Transitions

Besides these epistemological guidelines, this book also reviews and draws upon four bodies of literature which provide different perspectives and explanations on energy transitions, modernity and sustainability. These four bodies of literature are central to Chapter 2 of this book:

- *Socio-technical transitions* (or multi-level perspective) literature: uses a multi-level perspective to analyse transitions towards more sustainable socio-technical systems.
- *Ecological modernisation theory* literature: focuses on changing modes of governance and possibilities of radicalised modernity to overcome sustainability problems.
- *Energy practices* (or practice theory) literature: uses a practice-based approach to show how everyday life changes as a result of changing competences, meanings and materials.
- *Political ecology* literature: a framework for the investigation human-environment interactions at different scales, with reference to issues of political economy.

These four bodies of literature have been chosen because they represent some of the key social science approaches that have dealt with questions about energy, environment, development and sustainability over the last two decades. They are all well-established; most have been adopted by a number of core scholars and have featured in various key book-length reference works. In addition, they all draw, either implicitly or explicitly, on social theory and the core themes of this book, that is, modernity and sustainability. Moreover, because of their different disciplinary and ideological origins, contrasting them against each other helps to bring out some of the key tensions between modernity and sustainability. However, the aim is not to pigeon-hole certain (groups of) authors – indeed, other choices or groupings of literature would be possible – or pick winners, but to analyse the ways in which the respective authors deal with the key issues raised in this book, to what extent they help explain the empirical findings, and how they contribute to the formulation of a novel social scientific energy research agenda.

Towards a Critical Geography of Energy?

This book aims to contribute to what may be called a critical geography of energy, based on the extant theoretical and empirical discussions about energy, modernity and sustainability. The term 'energy geography', while by no means new, has been explored by Solomon et al. (2003), Pasqualetti (2011), Bouzarovski (2009a, 2009b) and Bridge et al. (2013). Pasqualetti (2011) states that only few publications continue to use the terms 'energy geography' or 'geography of energy' because 'the mix of geography and energy is so common it escapes casual notice' (p. 972). So, while it may seem to contradict the multi-disciplinary starting points outlined above, the word 'geography' in the 'critical geography of energy' is not so much included for its disciplinary connotation, but because questions about energy and modernity are deeply geographical (Bridge et al., 2013). Moreover, a critical geography would not necessarily be confined to the discipline of (human) geography. It could, and indeed should, be relevant to other disciplines.

Some of the key geographical elements related to questions of energy, modernity and sustainability have already been touched upon: the social construction and politics of scale, and the contested nature of sustainability. Other key examples explored in this book are the geography of cost and benefit (Hirsch, 1998) and the embeddedness of energy systems in social relations (Granovetter, 2002 [1985]). Finally, this research is based upon empirical or field-based research, which is in line with the epistemological framing of this research and following the messy reality of human-environment interactions. I consider this approach to research appropriate given the rapidly changing world of energy production and consumption and its associated growing environmental problems and inequalities. The following section outlines how the above aims, questions and epistemology translate into methodology.

Methodological Notes

This research adopts different methodologies for each level of analysis. Importantly, all of the chapters are framed as case studies, including those that look at Southeast Asia in general, Thailand and Laos on the country level, as well as the local energy trajectories. The key sources of information for the regional and country-level case studies (Chapter 3) are secondary sources of literature on the general history and specific to the development of the energy and power sectors. Where possible and relevant, primary data sources are used. These include reports, statistics and interviews with key stakeholders.

Local case studies were chosen because they have undergone (and continue to undergo) different forms of transition in their electricity production and consumption over the last two or three decades (see Figure 1.2 for the locations). Several field trips were undertaken to each of the four sites in 2010 and 2011. The key sources of data were qualitative interviews and observation, complemented by quantitative survey data and secondary sources of information.

Outline of the Book

Chapter 2 discusses and defines the three key concepts of this book: energy transition, modernity and sustainability. These concepts are then used to review four bodies of social science literature that broadly deal with socio-technical systems, energy and sustainability: socio-technical transitions, ecological modernisation theory, energy practices and political ecology. Acknowledging that each has a base in different disciplines and social theory, this chapter focuses on the different ways in which these bodies of literature relate and approach

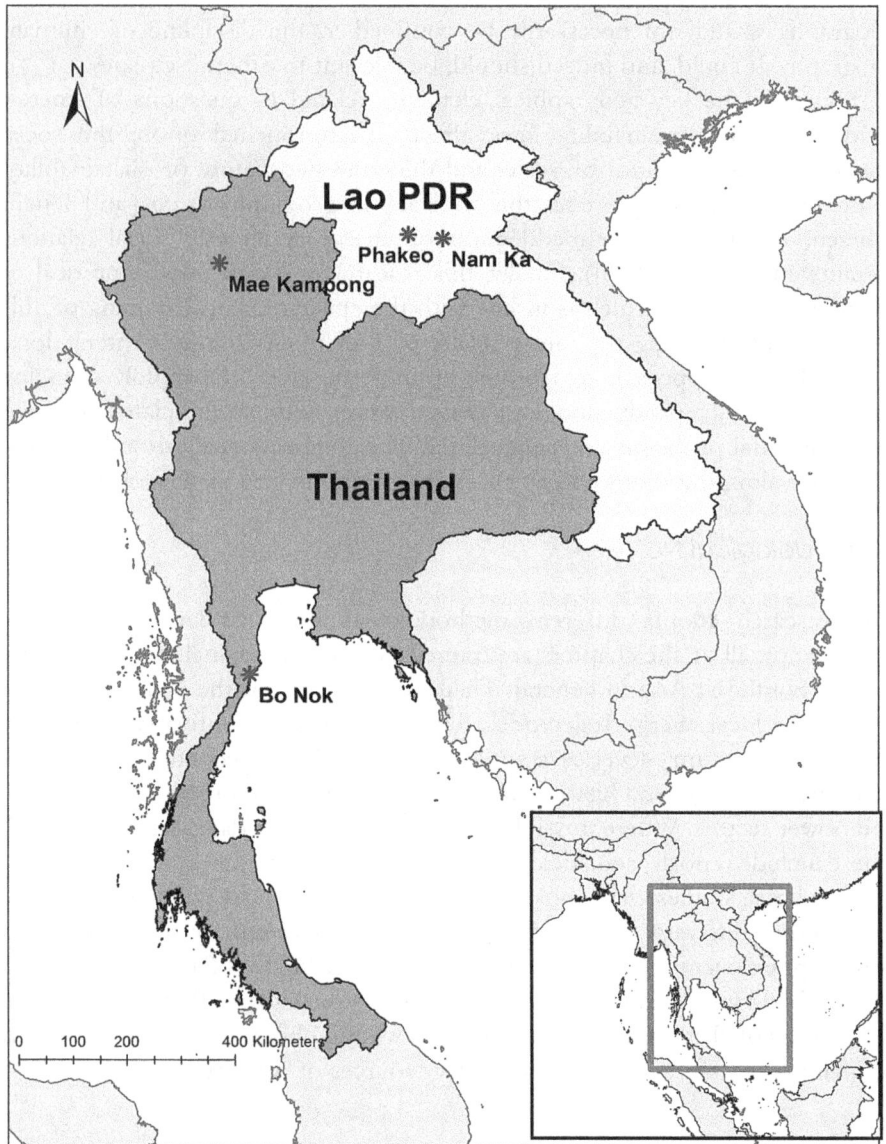

Figure 1.2   The location of the four case studies in Laos and Thailand
Source: Author.

the three key concepts. The last part of this chapter integrates these findings to argue that, whereas energy and modernity are dialectically related, sustainability should be seen as an integral part of this dialectic, rather than outside of it.

Chapter 3 analyses the development of state formation, territorialisation and the discourses of modernity related to energy and power sector development in

Southeast Asia, with a specific focus on Thailand and Laos. The first part of the chapter is an analysis of the regional neoliberal energy-modernity discourses in the plans of the Asian Development Bank and the Association of South East Asian Nations. The second part of the chapter zooms in on Thailand and Laos to show how energy development and modernity are intertwined at the national level as a tool of territorialisation, the creation of a common identity, and for the development of state-led projects – and to understand some of the differences, similarities, and the increasing mutual interdependence that obtains between the two countries.

Chapter 4 introduces the four case studies as energy trajectories, the aim being to show how they are embedded in the local histories, geographies and changing livelihoods. The chapter shows that these local case studies only reflect the national energy transitions to some extent, while diverging in other respects.

Chapter 5 analyses the data from household surveys and in-depth interviews to see how the energy trajectories play out for different people and their everyday practices in the four cases. Starting with the changing appliances and machines, the chapter then moves to the influence of the energy trajectories on different generations and the convergence of ethnic minority and majority cultures. The chapter ends with some observations about the changing experience of distance and time, the shift from community to individual, and the increasing invisibility of energy systems.

Chapter 6 discusses the findings from the three previous chapters in light of the contemporary academic debates (Chapter 2) and the three key concepts of this book. Moreover, it provides some programmatic notes towards the development of a critical energy geography, based on concepts such as embeddedness, the geography of cost and benefit, and competing sustainabilities. Moreover, it reflects on the use of energy transitions, energy trajectories, and normative positions in the study of energy, modernity and sustainability.

Chapter 7, the concluding chapter, presents the key issues discussed throughout the book, outlines some research and policy implications, and identifies avenues for further research into energy, modernity and sustainability in Southeast Asia and beyond.

Chapter 2
Energy Transitions, Modernity and Sustainability

Introduction and Key Arguments

Energy and its role in development and modernity is surprisingly understated in the social sciences despite its socio-economic importance and the challenges identified in the previous chapter, such as poverty, pollution, environmental degradation and climate change. This chapter reviews the extant social science literature that relates to the triad of energy, modernity and sustainability. Specifically, it focuses on four different bodies of literature: socio-technical transitions, Ecological Modernisation Theory, energy practices, and political ecology.

The first key argument of this chapter is that energy transitions have a critical role in shaping modernity, and, simultaneously, modernity in shaping energy transitions. In other words, this chapter argues that there is a dialectical relationship between modernity and energy transitions – the two elements are in a continuous state of flux wherein change in one affects change in the other – and neither takes primacy (Harvey, 1996). As such, this chapter critiques approaches that focus only on one dimension or one direction of this relationship, such as economic, technical or environmental aspects of energy transitions.

Another main argument is that environmental sustainability is a type of reflexive modernity, and therefore an integral part of the energy-modernity dialectic. It follows from this that approaches that treat sustainability as a trade-off or separate from this dialectic are likely to miss its embeddedness, to provide an incomplete understanding and, by extension, shallow policy recommendations.

Third, this chapter argues that there is an increasing momentum for social science to try to grapple with energy, modernity and sustainability. This book identifies limited or one-sided conceptualisations of modernity and weakly-developed frameworks as some of the key problems in the current literature. Therefore, this chapter presents an approach that advances a critical scalar analysis of energy transitions, modernity and sustainability that is empirically grounded throughout the rest of the book.

The rest of the chapter starts with a discussion of the three key concepts: energy transition, modernity and sustainability. Next, the (implicit) use of energy transitions, modernity and sustainability in the four bodies of social science literature is discussed to review the key positions and how they relate to debates surrounding the three key concepts. Subsequently, the chapter returns to the key concepts using the insights from the four bodies of literature.

Conceptualising Energy Transitions, Modernity and Sustainability

Energy Transitions

The terms 'transition' or 'energy transition' have different meanings in different (academic) contexts and are used to describe or analyse phenomena on different scales. According to the dictionary, the meaning of *transition* is '[a] passing or passage from one condition, action, or (rarely) place, to another' (Oxford English Dictionary, 2002b). It is not only a common word in everyday language, but also takes on particular meanings in different academic disciplines, can include different dimensions, causes and even be associated with different geographical areas, as shown in the next paragraph.

The academic use of transitions in the study of demographic transitions, agrarian transitions and transition economies show some of the common uses of the transitions, including the common assumption of linearity in transition models. Demographic transitions are understood as changes in the composition of the population, usually from one that has more of the younger age-groups, to one that has more of the older age-groups, circumstances attributable to changing birth and death rates (Lee, 2002). These kinds of transitions are about one focal variable, for example, age-structure, whereas the causes of the transition can vary. This is different from agrarian transitions, which deal with the changing composition of the agricultural economy, for example, from subsistence-based to market-based production (Goodman and Redclift, 1982). However, this use of transition usually includes elements such as rural-urban migration, the development of off-farm activities, and changing political relations (Rigg, 2001, 2003). In other words, in this case, the transition is intrinsically multi-dimensional as well as multi-causal. The term 'transition economies', finally, refers to countries moving from centrally-planned to free-market economies (Fischer and Sahay, 1996). In this example, the transition – while multi-dimensional and multi-causal – is associated with a particular set of countries and political ideologies.

In the context of energy, the term 'transition' has been used in different ways. Melosi (2006) states that '[i]n the broadest sense, the concept can help researchers understand the evolution of human material culture, economic growth and development, the utilization of resources and social organization. Utilized too narrowly however, it may merely provide a convenient instrument for segmenting energy history within a one-dimensional chronology' (p. 3). Most definitions are somewhere in between these extremes or focus on a particular aspect of energy transitions. As O'Connor (2010) suggests:

> Changes in the patterns of energy use take many forms. Energy resources include fossil fuels such as coal, oil, and natural gas, and renewable energy flows such as wind and solar energy. These are turned into energy carriers, such as electricity or gasoline. The carriers are then supplied to energy converters, such as a compact fluorescent light bulb or an automobile, and ultimately used to provide energy services such as lighting or transportation. An energy transition – a particularly significant set of changes to the patterns of energy use in a society – can affect any step in this chain, and will often affect multiple steps. (p. 8)

O'Connor's definition of 'energy transition' shows the complexities involved in thinking about energy chains and how transitions usually affect more than one link in said chains. Moreover, it challenges some models of energy transitions which assume an ever-increasing energy use and a teleological development toward 'modern' energy sources, inspired by Rostowian models of economic development (Rostow, 1960). One example of such a model is the household 'energy ladder', which is used by the influential International Energy Agency and by the United Nations. Figure 2.1 shows an example of such a ladder with different consecutive stages of energy levels, ranging from 'basic human needs' via 'productive uses' to 'modern society needs'. Such models assume an abundant supply of energy sources and that all households wish to move to the right side of the ladder. However, empirical evidence shows that the energy ladder often does not reflect the actual practice of energy transitions at household level (Hiemstra-van der Horst and Hovorka, 2008; Sovacool, 2012).

Besides these simple models, there is an increasing body of literature devoted to more sophisticated models of energy transitions, with which this book engages. An important predecessor to much of this work is *Networks of Power: Electrification in Western Society 1880–1930*, in which Hughes (1983) looks at the early developments of electricity systems in England, Germany and the United States using a scalar analysis. Hughes, an historian by training, combines micro-level analysis (for instance, Edison's early inventions and adaptation in the field of electricity) with macro-level events (such as the First World War) to explain the shaping of these 'large technical systems'. Hughes coins the term 'seamless web' (1986) to stress that social and material processes are highly intertwined and cannot be analysed in isolation, but should be approached as a 'socio-technical system'. This work also stresses the relation of energy transitions with modernity, a subject to which this chapter now turns.

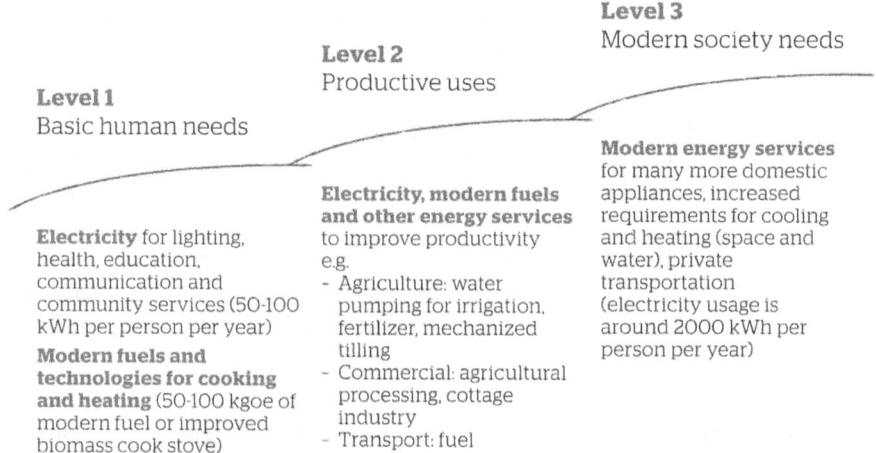

Level 1
Basic human needs

Level 2
Productive uses

Level 3
Modern society needs

Electricity for lighting, health, education, communication and community services (50-100 kWh per person per year)
Modern fuels and technologies for cooking and heating (50-100 kgoe of modern fuel or improved biomass cook stove)

Electricity, modern fuels and other energy services to improve productivity e.g.
- Agriculture: water pumping for irrigation, fertilizer, mechanized tilling
- Commercial: agricultural processing, cottage industry
- Transport: fuel

Modern energy services for many more domestic appliances, increased requirements for cooling and heating (space and water), private transportation (electricity usage is around 2000 kWh per person per year)

Figure 2.1 Example of an 'energy ladder'

Source: 'Energy for a Sustainable Future: Summary Report and Recommendations', by The Secretary-General's Advisory Group on Energy and Climate Change (AGECC), © (2010) United Nations..Reprinted with permission of the United Nations.

Modernity

The term 'modernity' is also used in many different ways, both in public discourse and academic disciplines (Martinelli, 2005). The term is derived from the Latin *'modus'* meaning 'now' (Delanty, 2007). According to the *Oxford English Dictionary* (2002a) it is '[a]n intellectual tendency or social perspective characterized by departure from or repudiation of traditional ideas, doctrines, and cultural values in favour of contemporary or radical values and beliefs'. Like transitions, it has different meanings for different scholarly disciplines. As such, it is also perhaps one of the most important, yet at the same time most contested, concepts in the social sciences, often intimately related to concepts such as capitalism, mobility, democracy and the development of the nation state (Strohmayer, 2009).

This book uses modernity in two different ways following Wagner's (2001) classification of modernity as 'an historical or empirical instance', or as 'a philosophical or epistemological condition'. The conceptualisation of modernity as an historical or empirical instance is related to the dictionary definition in the previous paragraph and is often used to describe the age of exploration and colonisation, the period of Enlightenment or the Industrial Revolution, depending on the author's position (Strohmayer, 2009). The terms 'modernisation' or 'modernisation theory' usually refer to this thinking in stages, as in the example of the energy ladder. This use of modernity will be referred to as 'modernity discourse' in this book. Some examples include reference to industrialisation or urbanisation; the use of 'expert' knowledge instead of 'traditional' sources of knowledge; thinking in terms of market economies rather than subsistence needs; the acceptance of political authority of the nation state rather than personal relations; and the construction of identity though mass media rather than through ancestor-based rituals (Rigg, 2007, pp. 59–60).

In contrast, modernity as a philosophical or epistemological condition is a reflexive mode of analysing people's changing knowledge of the world. One way of defining reflexivity is through the realisation that knowledge of the world changes the way one interacts with it. Giddens (1984) calls this the 'double hermeneutic'. The conceptualisation of modernity as a philosophical or epistemological condition is historically strongly linked to theories about the autonomy of human beings and the rise of rationality as the main way of ordering the world and one's knowledge of the world (Wagner, 2001). Such theories of modernity, often referred to as models of social change or simply as social theory, have become central to the discipline of sociology. However, in the past, this reflexive mode of modernity has often been conflated with classifications of modernity as an historical instance.

The distinction between modernity as historical instance or philosophical condition helps to understand its overlap with the term 'development'. It is similar to Hart's (2001) distinction between 'Big D' and 'little d' development:

'big D' Development defined as a post-second world war project of intervention in the 'third world' that emerged in the context of decolonization and the cold war, and 'little d' development or the development of capitalism as a geographically uneven, profoundly contradictory set of historical processes. (p. 650)

Following this distinction, modernity as historical instance – often referred to as modernisation – is closely related to 'Big D' development and to the way development is commonly referred to in Southeast Asia. By contrast, modernity as philosophical condition is more akin to 'little d' development. However, whereas Hart's discussion about D/development is heavily focused on the global South, modernity has a weaker geographical connotation. Moreover, as I will argue later in this chapter, I challenge the conceptualisation of modernity or 'little d' development as exclusively referring to capitalism.

The rise of post-modernism since the 1970s can be understood as a challenge to the conflation of modernity as a philosophical or epistemological condition and as an empirical or historical instance. Post-modernity is generally associated with developments in the field of linguistics, structuralism and post-structuralism, and associated with French authors such as Saussure,

Levi-Strauss, Foucault and Derrida (Palmer, 1997). Simply put, post-modernity rejects essentialisms (Derrida, 1972) because of the complexities, heterogeneity and inherent partiality of our understanding of the world. Moreover, post-modernity challenges the possibility of single narratives to explain development and social change. An important contribution of the debate surrounding post-modernity is that it revealed the contested nature and ambivalence of modernity. Moreover, it laid the foundations for the study of different interpretations and meanings of modernity; for example, through the study of discourse. Generally associated with the work of Foucault, discourse analysis is used 'to consider the way that we know what we know; where that information comes from; how it is produced and under what circumstances; whose interests it might serve; how it is possible to think differently' (Mills, 2003, p. 66).

The proliferation of post-modernity has led to a strong backlash and renewed interest in and theorisation about modernity since the mid-1980s (Knauft, 2002; Strohmayer, 2009). Wolfe (1998) argues that the literature on post-modernity – which challenged the dogmatic and Eurocentric aspects of modernity often in polemic and provocative ways – was in fact too successful and taken too seriously. This led to strong responses from philosophers and sociologists such as Jürgen Habermas and Zigmund Bauman, who challenged the idea of post-modernity by arguing that modernity itself is an unfinished project (Delanty, 2007). This chapter follows this tradition of authors who have attempted to come up with ways in which modernity may be preserved as an analytical tool. As such, it tries, on the one hand, to avoid falling into the trap of conflating historical and philosophical notions of modernity and, on the other hand, to avoid the implications of radical post-modernism, which refutes the idea of modernity altogether.

The next part briefly discusses three contemporary responses to theorising modernity that feature throughout this chapter. First, Bruno Latour (1993) has famously argued 'we had never been modern'. According to Latour, the Enlightenment view of reflexive knowledge and rationality has led to a false divide between nature and culture. This divide has been formalised through increasingly purified and fragmented domains of knowledge, as may be seen in academic disciplines, different sections of the news media, and different ministries. However, Latour argues, the more one tries try to maintain this divide, the more one is confronted with hybrid 'quasi-objects', such as environmental pollution, biotechnology and energy systems that do not fit on either side of this divide. This will eventually lead to the realisation that one's ideas about modernity are wrongly based on a separation between nature and culture, between humans and their environment, and between the technical and the social, to name just a few.

Another response in the wake of post-modernity is the idea of reflexive modernity – also referred to as second or late modernity – which was developed by two influential sociologists, Anthony Giddens and Ulrich Beck (Beck, 1992;

Beck et al., 1994). While Beck and Giddens agree with post-modernists that there has been a 'break' with first or simple modernity, they argue for more or radicalised modernity rather than turning away from it. In the words of Beck et al. (2003), '[w]hen modernization reaches a certain stage it radicalizes itself. It begins to transform, for a second time, not only the key institutions but also the very principles of society. But this time the principles and institutions being transformed are those of modern society' (p. 1). Understood in this way, reflexive modernity thus refers to a meta-change in society, comparable with the changes that have led to 'first modernity'. Beck et al. (2003) argue that the drivers of this change are increasing globalisation, intensification of individualisation and related transformation of gender roles, a breakdown of the full employment society, and the perception of a global ecological crisis.

A final set of responses may be summarised as 'alternative modernities'. They are mainly associated with cultural theorists and anthropologists, including many scholars from the global South (Chakrabarty, 2000; Eisenstadt, 2000; Gaonkar, 2001; Knauft, 2002). Like reflexive modernisation, the idea of alternative modernities does not see the 'end' of modernity, but rather multiple (non-Eurocentric) expressions of modernity and tradition, which lead to different outcomes in different contexts. This approach challenges the notion that modernity necessarily leads to convergence in the form of societal modernisation, such as the spread of market economies, the nation state and individualisation (Gaonkar, 1999). It also stresses 'the ambivalence of modernity, which cannot be reduced to a single dimension' (Delanty, 2007, pp. 3,069–70). Thinking in terms of alternative modernities also opens up the opportunity to link anthropology with the global political economy in ways that do not privilege any particular scale or causal directionality (cf. Tsing, 2005). Alternative modernity or alternatives to modernity are often referred to in debates about environmental sustainability.

Sustainability

The last key concept of this chapter is 'sustainability', which in this book, is referred to as 'environmental sustainability'. Marsden (2009) argues that sustainability became a 'conceptual question' (p. 11: 103) in the context of increasingly pressing environmental problems and crises related to the capitalist mode of production. However, its popular use stems from its use in conjunction with development – sustainable development – and its iconic definition in the report titled *Our Common Future* (Brundtland, 1987). Since then, the interpretation of sustainability has changed with time. Nevertheless, as Prudham (2009) argues, there are 'good reasons to take this word [sustainability] and some of what it conveys seriously, and in particular to differentiate the word from the term "sustainable development"' (p. 737).

One of the key problems of sustainability is that it is often treated in an instrumentalist and reductionist way, principally by policy makers, economists and managers (Prudham, 2009). In this way, sustainability often becomes a narrowly defined way of seeing pollution, emissions or waste as opposed to development and modernity. This is why the concept of 'sustainable development' often becomes used in an oxymoronic sense, a contradiction in terms devoid of any meaning (Redclift, 2005). This chapter argues that there are ways in which the term 'sustainability' may be understood more meaningfully, in conjunction with energy transitions and modernity in four bodies of literature. To this end, this chapter supports the argument that 'an important intellectual trend has been to recognize that sustainability theory needs to integrate and be critically interwoven with broader and challenging spatial theories of the economy and polity' (Marsden, 2009, p. 11: 107). This is the subject addressed in the next part of this chapter.

Overview of Social Science Approaches to Energy Transitions

Having introduced the three key concepts separately in the previous section, the second part of the chapter relates them to four established bodies of social science literature: socio-technical transitions, Ecological Modernisation Theory, energy practices and political ecology. The choice of literature and approaches represents a deliberate selection from the social science literature which touches upon energy transitions, modernity and sustainability (see Table 2.1 for some key features). The first reason for selecting them is that each is focused broadly on the linkages between society, environment and technology from a holistic perspective. Second, all of the approaches are broadly concerned with issues related to sustainability. Third, each approach engages with social theory and modernity, albeit some more explicitly than others. An important difference, however, is that the emphasis of each of the bodies of literature is on different aspects or scales of energy transitions, for example, production or consumption, specific technologies or systems, and different scales of analysis. As such, they are not necessarily opposed to each other, and sometimes they are complementary. Moreover, the boundaries drawn around the different bodies of literature may be drawn differently and on occasion there is some overlap.

The structure of each of the sections dealing with each of the four bodies of literature is as follows: first, they are each introduced briefly and examples related to energy transitions are presented, with special attention to cases in Southeast Asia. Next, the key social theories underpinning each body of literature are discussed, with specific reference to the conceptualisation of modernity and sustainability. The final part of each section assesses the approaches on their own merit and seeks to determine how they might contribute to a social scientific approach to energy transitions in Southeast Asia.

Table 2.1 Four approaches to understanding energy and social change

	Socio-technical transitions	Ecological Modernisation Theory	Energy practices	Political ecology
Related academic disciplines	STS, innovation studies, economics	Environmental sociology, political science	Sociology, Cultural Theory, STS	Human geography, human ecology, anthropology
Selected key authors	Rip, Kemp, Geels, Kern, Grin, Schot, Raven	Mol, Buttel, Spaargaren, Hajer, Jänicke, Sonnenfeld	Shove, Warde, Walker, Pantzar, Schatzki	Watts, Peet, Robbins, Neumann, Forsyth, Bridge
Issues (examples)	Historical transitions, development of new technologies, niche and transition management	Environmental regulation, state reform, environmental governance, social movements	Everyday routines, normalisation of practices	Conflicts over natural resources, neoliberalism, scale
Basis in social sciences	Structuration theory (Giddens), evolutionary economics, complexity theory, ANT	Structuration (Giddens), Risk society (Beck), networks, flows and mobilities (Castells and Urry)	Theory of practice (Bourdieu, Schatzki, Reckwitz), Structuration (Giddens)	Neo-Marxist political economy, (post-)structuralism, knowledge/power (Foucault)
Conceptualisation of modernity	No explicit concept of modernity	Reflexive modernity, green capitalism	Modernity through practice, routines, and normalisation	Modernity as capitalism, modernity as discourse
Conceptualisation of energy transitions	Multi-level perspective on socio-technical change, path dependency	Energy transitions as part of environmental reform	Transitions as changing (sets of) practices	Energy transitions as a contested (capitalist) project
Conceptualisation of sustainability	Sustainability can become mainstream through 'management' of niches of sustainable innovations	Sustainability can be institutionalised through reflexive modernisation	Sustainability is integral part of a system of energy practices/routines	Sustainability as a contested and multi-scale discourse
Ontology	'evolution theory and interpretivism' (Geels, 2010)	Reflexive modernisation	Practices, flat (relational) ontology	Critical realist, (post-) structuralist
Key methods	Historical case studies, systems analysis, transition management	Sectoral/national case studies of industries and practices	Case studies of practices, ethnography	Ethnography, political economy, discourse analysis
Units of analysis	Technologies (innovations), socio-technical systems	Sectors, industries, institutions, practices	Practices	Human-environment interaction, local communities, institutions

Source: Author.

Socio-Technical Transitions (Multi-Level Perspective)

Introduction and Conceptualisation of Transitions

The socio-technical transitions literature is a well-established 'mid-range theory' (Geels, 2007), which aims to explain how long-term transitions take place, specifically those towards more 'sustainable' systems (Grin et al., 2010). While the emphasis of this literature is on technologies and technological innovation, it also provides tools to analyse how technologies are embedded in institutions, rules and systems. The term 'socio-technical system' which was coined by scholars in the field of Science Technology and Society (STS), and is commonly associated with the pioneering works collected in the edited volume *The Social Construction of Technological Systems: New Directions in the Sociology and History of Technology* by Bijker et al. (1987). This volume is noteworthy for having brought three distinct fields of study together. The first, a historical approach to the study of technological systems related to the earlier mentioned work of Hughes (1983) was later consolidated in the large technical systems (LTS) approach, the focus of which was on large scale infrastructural works, such as electricity, water, sanitation and waste systems (Coutard et al., 2005; Mayntz and Hughes, 1988; Summerton, 1994). The second approach that emerged from systems-thinking, was actor-network theory (ANT), which emphasised the hybrid nature of innovations by stressing the role of human and non-human 'actants'. This philosophy has since gained wide currency in the social sciences, well beyond STS (Latour, 2005; Law, 1992; Law and Hassard, 1999). The third approach, the social construction of technology (SCOT), stresses the evolutionary nature and social shaping of technical systems, using terminologies such as 'path dependency', 'lock-in' and 'interpretive flexibility'. While the terms SCOT and LTS have now largely disappeared, and ANT has taken on a direction of its own, the socio-technical transitions literature owes much to these early developments in STS. In many ways, it is a continuation of the themes and issues discussed in these three approaches (Geels and Schot, 2010, p. 33).

There are a number of 'schools' in the socio-technical transitions literature, each with their own emphasis. The main schools are the multi-level perspective (MLP) and the technological innovation systems (TIS) approach, both of which originated in the Netherlands. The former focuses on individual innovations and how they succeed or fail when dealing with the extant socio-technical regimes: the latter has a more macro-level institutional outlook (Markard and Truffer, 2008). While there is substantial cross-over in terms of social theory and scholars in both approaches, this book will focus on the MLP and less on the TIS approach, because the former has a stronger academic base (publications, networks) and relates more to energy transitions. The Viennese socio-metabolic transitions approach (Fischer-Kowalski and Haberl, 2007; Fischer-Kowalski and Rotmans, 2009) will also be excluded, due to its different roots – in human ecology and environmental history – and its material, long-term and macro-level focus.

The multi-level perspective of socio-technical change can be summarised by the model in Figure 2.2, which is reproduced in many publications dealing with this approach. This model, which was introduced by Rip and Kemp (1998) and further developed by Frank Geels (2002, 2005), shows the different 'levels' of socio-technical system – the niche, the regime and the landscape – and their interactions. The key level in the model is the socio-technical regime, which consists of the rules and routines embedded in infrastructure, markets, technology, policies, knowledge and meaning. The model of change is mainly based on the bottom-level, the technological niches. Socio-technical transitions scholars claim that technological niches challenge the regime with its established institutions and rules. Should these niche technologies 'succeed', they may ultimately alter the parameters of the regime and lead to long-lasting changes (Kemp et al., 1998; Schot and Geels, 2008). The top-level, the landscape developments, includes broad macro-level changes beyond the influence of the other two levels, for instance, climate, industrialisation, wars and globalisation (Geels and Schot, 2010). While the MLP has primarily been

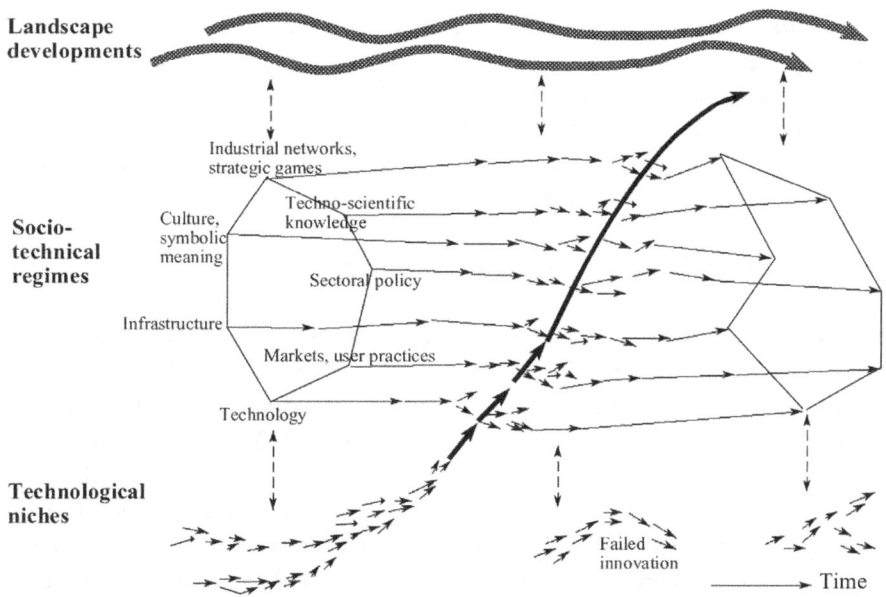

Figure 2.2   Diagram illustrating the multi-level perspective on socio-technical transitions

Source: Reprinted from Geels (2002, p. 1,263) with permission from Elsevier.

23

used to analyse historical cases of transitions, it has also been used to provide recommendations for the 'management' and governance of transitions (Kemp et al., 2007a; Loorbach, 2010).

Many case studies in the socio-technical transitions literature deal with energy transitions. Both the model and the case studies using the approach are rooted in empirical understandings of energy transitions and a commitment to historical analyses as well as policy engagement. Prominent examples are studies explaining Dutch energy transitions from coal to gas and the efforts to introduce renewable and decentralised energy (Raven, 2007; Verbong and Geels, 2007; Verbong et al., 2008). The Dutch government's attempt to manage transitions as a direct outcome of this literature has yielded a large number of articles reflecting on this 'experiment' (Hendriks, 2008; Kemp, 2010; Kemp et al., 2007b; Kern and Smith, 2008; Smith and Kern, 2009). While there are plenty of cases outside of the Netherlands, most of them are still confined to Western Europe. Only recently, there has been evidence of increased attention to sustainability transitions in Asia. For example, a special issue of *Technological Forecasting and Social Change* in 2009 2volume 76, issue 2) posed to the question 'Sustainability Transitions in Developing Asia: Are Alternative Development Pathways Likely?', followed by an article by Berkhout et al. (2010) on sustainability experiments in Asia. In these articles, the authors propose their model of socio-technical systems as an alternative to the 'standard' economic growth models and (environmental) Kuznets curves in the newly industrialising and emerging economies in Asia. Their methodology focuses on institutional learning, sustainability experiments and horizontal and vertical linkages. Among the very few articles that apply the MLP to Southeast Asia are articles by Hansen and Nygaard (2013, 2014) on the palm oil biomass waste-to-energy niche in Malaysia and its international linkages.

Engagement with Social Theory and Modernity

The socio-technical transitions literature draws only sparingly on social theory, and it has a very limited conceptualisation of modernity. Socio-technical transitions scholars mainly refer to Anthony Giddens's social theory, in particular to his work on structuration (Giddens, 1984). Several authors have used this work to explain the dynamics and interaction between the different levels of analysis, between niches and regimes, and between regimes and the landscape. More specifically, they have used it to explain how actors are not only embedded in certain structures, but reproduce these same structures. In other words, the scholars in this field are trying to preserve a focus on individual agency, which is particularly important for niche-level analyses, while at the same time justifying the need to look at the broader structures that shape the long-term transitions (Geels, 2004; Geels and Schot, 2010). Besides Giddens, there are also references to the work of Bourdieu and his concept of 'habitus' as a way of overcoming the structure-agency divide (Grin, 2010). Yet, these comments are even less well articulated.

The multi-level perspective (MLP) also uses Beck and Giddens's notion of reflexive modernisation, mainly to show the contingency and complexity of transitions and governance. Grin (2010), for example, argues that 'modernisation theory' helps to overcome simplifications in debates about governance, in particular relating to the separation of state, market, society and science, and the diminishing role of government in the process of governance. Instead, Grin argues for a more complex view of governance, drawing upon Beck's concept of 'second modernity'. According to Grin, this shows that transitions to sustainable development have to be understood as contested and as taking place in a pluralistic society, wherein actors are constantly adjusting their positions and are aware of the 'politicisation of side effects' (p. 230). Elsewhere, Rotmans and Kemp (2008) claim that '[t]ransition management is thus a strategy of reflexive modernization. It accepts that there may be risks and rebound effects from system innovations, which must be anticipated and countered' (p. 1,010).

In an explicit attempt to engage with social theory, Geels (2010) positions the multi-level perspective in relation to the following seven 'ontologies': rational choice, evolution theory, structuralism, interpretivism, functionalism, conflict and power struggle, and relationism. He argues that not only is the multi-level perspective mainly based on a crossover of evolution theory and interpretivism, but that it also selectively draws on ontologies of conflict and power struggle, and structuralism. Unfortunately, Geel's goal of discussing seven 'ontologies' in one article did not allow for much elaboration of each. Rather, the article pigeon-holes social theory to fit the multi-level perspective and strips it of its complexities and nuances.

Assessment and Relevance of Socio-Technical Transitions Literature for this Book

The socio-technical transitions approach is a multi-disciplinary body of literature which has built up a considerable academic base consisting of case studies drawn from various countries, related to energy and other sustainability transitions. One of the key strengths is the empirically and historically informed model of how transitions unfold and how they are embedded within social norms, institutions, and existing socio-technical regimes. Another noteworthy aspect is how in recent years scholars involved in this approach have extended their investigation to social theory (Geels, 2010), spatial and geographical issues (Coenen et al., 2012; Raven et al., 2012; Späth and Rohracher, 2014) and other geographical regions, such as Asia (Berkhout et al., 2009; Berkhout et al., 2010). In terms of this research, the socio-technical transitions approach brings a number of empirical insights in energy transitions to the table. The work should also be credited for its attempt to translate the theory into policy intervention strategies.

Nevertheless, there are a number of limitations of the socio-technical transitions approach. First, its focus on innovations and technologies is often at the expense of other factors that play a role in the transition towards sustainability. One of the problems is the limited conceptualisation of power and politics (Meadowcroft, 2009, 2011) and the lack of grounding of transitions in geographical contexts of uneven development (Lawhon and Murphy, 2011). While some of these points are addressed to some extent in recent contributions, it is unclear whether the approach will be flexible enough to accommodate these issues properly in the future.

A more fundamental problem is the unclear and sometimes contradictory conceptualisation of sustainability, due in the main to the Janus-faced tensions between the academic and normative aspects of the literature. For example, in their introduction to their work *Transitions to Sustainable Development: New Directions in the Study of Long Term Transformative Change*, Grin et al. (2010) initially talk about sustainable development as 'an open-ended orientation for change' (p. 2). Then, two paragraphs later, they advocate 'shaping transitions towards a specific normative orientation – in this case, sustainable development' (p. 3). In another article, Geels (2011) maintains that 'sustainability transitions are goal-oriented or "purposive"' and that 'sustainability is a collective good'. However, in the same paragraph, he further notes that '[b]ecause sustainability is an ambiguous and contested concept, there will be disagreement and debate about the directionality of sustainability transitions (Stirling, 2009), the (dis)advantages of particular solutions and the most appropriate policy instruments or packages' (p. 25).

In other words, there is a tension in the use of sustainability, which is sometimes conceptualised as fundamentally open-ended and contested, and at other times as normative and goal-oriented. Especially in the latter case, sustainability is treated as something external and disconnected from the messy reality of socio-technical development, rather than as something internal to it. Shove and Walker have articulated this problem in a number of places (Shove and Walker, 2007, 2008, 2010; Walker and Shove, 2007). They point out that most of the socio-technical transitions literature subscribes to the notion that sustainability transitions can be brought about by generating consensus and creating a rational path towards sustainability. However, it is unclear what this path is and who gets to decide.[1] Moreover, they show that this is particularly problematic where the socio-technical transitions model is applied in practice

---

1 Later examples of this book show that the tensions between normative and open-ended conceptualisations of sustainability and the problem of governance between different scales and generations are not unique to the socio-technical transitions literature.

through 'transition management' (cf. Scrase and Smith, 2009; Smith and Stirling, 2010).

Another fundamental concern is the limited conceptualisation of modernity and social theory in the socio-technical transitions literature, in particular because it claims to be about *socio*-technical transitions. The review of the literature shows that only scant attempts have been made to engage with social theory, mainly by (token) references to Beck and Giddens. Where issues of modernity are debated, such as the influence of global capitalism, this is usually considered to be part of the 'landscape' level, which, Geels (2011) admits, is currently somewhat of a 'residual analytical category' (p. 36). In response, one might argue that the MLP is explicitly framed as a 'middle range theory', located in between empirical generalisations and 'grand theories' (Geels, 2007). However, given that the proponents of the MLP are interested to understand and sometimes even 'manage' socio-technical transitions, one would expect more explicit theorising of the social, in addition to historical analyses of technological innovations. Without theories of how technological innovations might affect the relations between the state and its citizens, the experiences of modernity, and the influence of technological changes on peoples' livelihoods, there is a danger of an overly technocratic approach (Lawhon and Murphy, 2011) and a democratic deficit (Hendriks, 2009). This point also echoes Shove and Walker's (2007) argument that 'for all the talk of socio-technical co-evolution, there is almost no reference to the ways of living or to the patterns of demand implied in what remain largely technological templates for the future' (p. 768).

In sum, while the socio-technical transitions literature (here, mainly the MLP) has been important in putting the study of energy transitions on the social science agenda, it lacks the depth to facilitate an understanding of the tensions between modernity and sustainability of energy transitions. This chapter now turns to a body of literature which attempts to engage with these issues more explicitly.

Ecological Modernisation Theory

Introduction and Conceptualisation of Transitions

Ecological Modernisation Theory (EMT), which is based in the disciplines of environmental sociology and political and environmental science, was developed as a response to environmental problems and its subsequent reaction in policy and social movements in the 1970s and 1980s (Mol, 1997). The literature mainly focuses on specific industrial sectors and institutional environments. The two most important ways in which ecological modernisation has been used – as distinguishes by Buttel (2000) – include the 'social-constructivist ecological modernisation' (cf. Hajer, 1995; Dryzek, 2005) and 'ecological modernisationist/ sociological' work, inspired by German industrial ecology and centred around the works of Dutch academics Arthur Mol and Gert Spaargaren (Mol, 1997, 2010a; Mol and Sonnenfeld, 2000; Spaargaren et al., 2006b). In this book, I will focus on the latter strand of EMT, which has arguably been more developed over the years as a distinct body of literature and has actively engaged with social theory and modernity.

The basic idea of ecological modernisation, according to Arthur Mol (2010a), is as follows: 'at the end of the second millennium, modern societies witness a centripetal movement of ecological interests, ideas and considerations in their institutional design. This development crystallizes in a constant ecological restructuring of modernity' (p. 66). In other words, Mol argues that societal change towards ecological concerns leads to changing institutions, environmental reforms, and eventually culminates in more environmental sustainability. Most of the EMT literature shows the extent to which certain sectors and industries are moving towards this reform. It also discusses the changing roles of the state, private sector, civil society, and processes such as globalisation.

Similar to the literature on socio-technical systems, EMT remains a largely Western-oriented body of literature, although it also features case studies from other parts of the world (e.g. Mol and Sonnenfeld, 2000; Mol et al., 2009; Spaargaren et al., 2006b). There is a paucity of case studies on the energy industry or reforms of electricity sectors, despite the availability of potentially rich material. Among the exceptions is a book by Toke (2011), who employs an EMT-based framework to study renewable energy industries in the USA, Germany, Spain, the UK, Australia and China. Toke seeks to ascertain why renewable energy has or has not been taken up in these countries. Another example is a study by Breukers and Wolsink (2007) on the slow development of the on-shore wind energy industry in the Netherlands. This tardiness, they suggest, is due to the lack of institutional capacity building. Finally, a PhD dissertation by Van Vliet (2002) discusses the relations between citizen-consumers and 'providers' in the development of innovations in the Dutch water and electricity networks. EMT studies on Southeast Asia are limited to non-energy specific studies, such as case studies in Thailand on the pulp and paper (Sonnenfeld, 2000), the dairy, and the appliance industry (Thongplew et al., 2013; Thongplew et al., 2014) and national-level studies, such as on Vietnam (Frijns et al., 2000).

Engagement with Social Theory and Modernity

Ecological Modernisation Theory not only claims the predicate 'theory', but also aspires to become a social theory, as article titles *Ecological Modernization as Social Theory* (Buttel, 2000) and *Ecological Modernization as a Social Theory of*

Environmental Reform (Mol, 2010a) suggest (see also Buttel et al., 2006). The position of EMT vis-à-vis social theory can best be explained as a response to the policies and protests originating the 1970s and 1980s, which, in turn, arose in the context of the global awareness of environmental crises (including the oil crises), and increasing public awareness of global environmental degradation and the limits to growth. At the time, many of these responses were framed as neo-Marxist critiques, linking the expansion of global capitalism to environmental degradation, through approaches such as the treadmill of environmental degradation (Schnaiberg, 1980, 1997), world systems theory (Wallerstein, 1974), and political ecology (see next section). EMT, critiquing these critiques, argued that environmental reforms were possible within the current social order, which includes capitalism (York et al., 2010).

The reasons why EMT scholars engage in discussions about modernity are set out in Spaargaren and Mol's (1992) work *Sociology, Environment, and Modernity: Ecological Modernization as a Theory of Social Change*, in which they argue that 'environmental problems are not just the unintended consequences of an otherwise fortuitous trajectory of modernity. These problems appear increasingly bound up with modernity in such a fundamental and "organic" way that they cannot be dealt with in isolation from it' (p. 323). In order to substantiate their position, the 'second' generation of EMT drew heavily upon the work of Ulrich Beck (1992) and (to a lesser extent) Anthony Giddens and their notions of reflexive modernisation (Beck et al., 1994). However, where Beck and Giddens see reflexive modernity as a distinctly different phase, EMT interprets it as a continuation of the current social order. As Buttel (2000) notes:

> [a]n ecological modernization perspective hypothesizes that while the most challenging environmental problems of this century and the next have (or will have) been caused by modernization and industrialization, their solutions must necessary lie in more – rather than less – modernization and 'superindustrialization'. (p. 61)

More recently, EMT has been moving in a different direction with regard to social theory. The aim was to incorporate Castells's (1996) and Urry's (2000, 2003) post-modernist approaches, in which networks and flows replace 'traditional' social science concepts such as society and the state (Mol, 2010a, 2010b). The 'third generation' of EMT has taken up these ideas, most notably in Spaargaren et al.'s *Governing Environmental Flows* (2006b), but also by tackling issues such as globalisation (Mol, 2001). Environmental flows are understood by the authors as material flows such as energy, water and waste, as well as governance arrangements and networks supporting and directing these flows (Spaargaren et al., 2006a). Moreover, this is seen by these leading EMT scholars as a way to incorporate ideas about post-modernity into their framework of reflexive modernity. However, a quote by Buttel et al. (2006) shows that the central premise of EMT – modern industrialisation is moving towards more sustainability through reflexive modernity – remains unchanged: '[w]hile ecological modernization is solidly modernist – albeit "reflexively modernist" – in its ontology, the environmental flows perspective tries to confront some of the "postmodern" challenges as discussed by Urry and Castells' (p. 358).

Assessment and Relevance of Ecological Modernisation Theory Literature for this Book

Ecological Modernisation Theory represents an attempt to link modernity and environmental governance. While drawing upon some of the same social theorists as the socio-technical transitions approach – Beck and Giddens – EMT scholars locate their input at the heart of understanding the interplay between social change and environmental transitions. One of the strengths of EMT is that it draws attention to the variation between different industries, firms and sectors. This would make it possible to move between scales and not start with a priori judgements of certain processes (York et al., 2010). Despite this strength, EMT has received limited attention from geographers, with the exception of a special issue in *Geoforum* (2000, volume 31, issue 1) and articles by Christoff (1996) and Gibbs (2000, 2006). Finally, as Watts (2000) observes, the insights of Ecological Modernisation Theory, and its focus on 'First World' issues of politics and the environment, may complement approaches such as political ecology, which focuses mainly on the global South.

A critical point is the way in which the idea of reflexive modernity is employed by EMT scholars, because it differs from the way it appears in Beck and Giddens's original concept. As Buttel (2000) points out, Beck draws a sharp line between first and second/reflexive modernity (Beck et al., 2003), and emphasises a new types of politics, which he called sub-politics (Beck, 1992). In contrast, EMT neither makes such a distinction between first and second/reflexive modernity, nor engages with the idea of sub-politics. Moreover, while EMT provides evidence of cases where reflexive modernity as increased industrialisation – in response to the problem created by capitalism and industrialisation – works, this does not necessarily mean that it will work everywhere and in all circumstances (York et al., 2010). Moreover, as Warner (2010) points out, '[w]hen solutions are framed in the same terms as the problems, the scope for critical theory, democratic politics and reflexive thinking is narrowed' (p. 551).

This diverging interpretation of reflexive modernity has important consequences for environmental sustainability. Reflexive modernity, if applied according to the way in which EMT scholars currently use it, can too easily result in superficial implementation of sustainability or 'greenwashing' (Baker, 2007). Indeed, as Langhelle (2000) argues, '[e]cological modernization should be seen as a necessary, but not sufficient, condition for sustainable development' (p. 303). In other words, while there is some evidence linking ecological modernisation and reflexive modernisation, this needs to be done more radically and taken well beyond the current EMT focus. Moreover, and following on from the previous point, more emphasis should be placed on ecological modernisation as a discourse rather than as an apolitical process that occurs outside the interests and agency of different actors with different (political) motives (Gibbs, 2006).

However, it would be unwise to dismiss the insights of EMT altogether. Fisher and Freudenburg (2001), for example, note that 'disagreements between ecological modernization and earlier bodies of thought, the debates over ecological modernization, to date, have too often been expressed in terms of black/white differences or extremes' (p. 704). Recently, various criticisms have been addressed by EMT scholars, in particular the challenge of technological determinism, its productivist focus and the neglect of aspects of consumption (Mol, 2010a; Mol et al., 2014). Spaargaren (2011), for example, has attempted to situate the argument surrounding ecological modernisation into a practice-based framework, which aims to overcome both the pitfalls of structuralist approaches and those focused on individuals (cf. the structure-agency debate). It is to this body of literature – practice theory or energy practices – that the chapter now turns.

Energy Practices (Practice Theory)

Introduction and Conceptualisation of Transitions

The 'energy practices' literature, which is rooted in practice theory, provides a different perspective on energy transitions and modernity. The emphasis of this literature is on the household and consumption scale of transitions, but there are other and more significant differences. While theories of practice are manifold in the social sciences, their application in the field of energy, technology, environment and sustainability is relatively new albeit growing in volume. The idea of thinking about practices can be traced back to philosophers such as Wittgenstein and Heidegger (Shove et al., 2012), although more contemporary scholars of practice draw on work by authors such as Bourdieu, Giddens, Foucault, Garfinkel, Latour, Butler and Taylor (Reckwitz, 2002). In a widely cited paper, Reckwitz (2002) provides the following definition of practice:

'practice' (Praktik) is a routinized type of behaviour which consists of several elements, interconnected to one other: forms of bodily activities, forms of mental activities, 'things' and their use, a background knowledge in the form of understanding, know-how, states of emotion and motivational knowledge. (Reckwitz, 2002, p. 249)

The establishment of social practice as the key unit of analysis aims to avoid having to privilege the individual (agency) or society and other forms of organisation (structure). This is because approaches focusing on the individual are often unable to account for larger scale phenomena. Similarly, approaches focusing on structure often lack clear conceptualisation of change in individuals. Giddens's (1984) structuration theory is a key inspiration for the practice-based approach discussed here; the basic idea being that activities should neither be understood as fully the result of human action nor as completely determined by social structures. Instead, focus should be upon the different components which constitute practices as an intermediate level, structuring reality temporarily (Shove et al., 2012).

In order to differentiate it from the wide range of other topics that practice theory could be applied to, this chapter focuses practice theory that was developed by a group of scholars for the purpose of raising questions about technological change and sustainability, questions often related to energy production and consumption (Shove and Walker, 2014). This is because, according to Spaargaren (2011), 'the first generation of practice theories as developed by Giddens and Bourdieu had little to offer for analysing the role of objects, technological systems, and hybrids' (p. 817). Thus, rather than asking questions about technological innovations and institutions, energy practice scholars subscribe to the notion that 'people do not really "consume" energy. Instead, they consume the services – heating, lighting, showering, cooking, television watching, computer interaction – that infrastructure of gas, electricity and water make possible' (Shove, 2004a, p. 1,054). This also shows how a practice-based approach focuses on everyday routines, and uses this to criticise top-down and determinist approaches to socio-technical change.

The application of practice theory to energy transitions is mainly found in fields such as the sociology of consumption (Wilk, 2002) as well as in economic sociology (Biggart and Lutzenhiser, 2007). The work of Elizabeth Shove has been pivotal to the establishment of energy practices literature and its grounding in social theory. And while the scope of this project becomes clear in some of her earlier publications (Shove, 1997; Shove et al., 1998), the key arguments may be found in her book titled *Comfort, Cleanliness and Convenience: The Social Organization of Normality* (Shove, 2003a). In this volume – which is a compilation of case studies of 'mundane' practices such as bathing and showering, doing the laundry, freezing, cooling, and air-conditioning – she shows how these everyday practices are related to material developments (such as technology), changing values and meanings, and shifting competences (Shove, 2003b). One of her key arguments is the non-linearity and 'horizontal' converging of

32

these practices across cultures and societies, rather than the 'vertical' (multi-level) dissemination emphasised in the socio-technical transition literature. In other words, rather than focusing on new innovations that emerge in niches which challenge regimes, energy practices literature emphasises how changes in routine behaviour are at the core of socio-technical transitions. Moreover, these practices now travel and change faster and more easily through the proliferation of transnational companies, and the cross-cultural influences of international travel and media (Shove and Pantzar, 2005; Shove et al., 2013). This change in practices has important implications for the use of energy and other resources and environmental sustainability. As argued by (Shove, 2003a):

> [t]he real environmental risk is not that services will be redefined (this happens all the time), but that there will be sweeping, cross cultural convergence in what people take to be normal ways of life, and a consequent locking in of demand for the resources on which these ways depend. (p. 199)

At present, there is limited literature using energy practices to discuss energy issues in Southeast Asia or other countries in the global South. The exceptions are a series of articles on thermal comfort, air-conditioning and Western influences on household energy consumption patterns in the Philippines (Andamon et al., 2006; Sahakian, 2010, 2014; Sahakian and Steinberger, 2011). In addition, there is Wilhite's (2008, 2012) research into transformations of everyday life in India with particular focus on changing energy and water consumption practices. A final example is a book chapter by Paling and Winter (2011), who discuss the household-level impact of rapid economic development and images of modernity on environmental sustainability in Cambodia. These publications demonstrate that the energy practices framework has great potential for analysing the changing practices which emerge as a result of the interaction between existing local practices and the partial and dispersed modernity in Southeast Asia and the global South.

Engagement with Social theory and Modernity

Because energy practices constitute part of a wider set of practice theories, there is ample reference to social theory, as already suggested above. The conceptualisation of modernity underpinning or following on from energy practices is complex given that it is not a central proposition within energy practices like in Ecological Modernisation Theory. The centrality of the issue of modernity and social change may be seen in the first line of *The Dynamics of Social Practice* which asks: 'How do societies change? Why do they stay so much the same?' (Shove et al., 2012, p. 1). Shove's commitment to the issues and dilemmas of modernity may also be seen in the following questions:

How do practices, expectations and ways of life become naturalized? What energizes processes of escalation, standardization, differentiation and development? In addition, how do models, ideologies, cultures and ways of life spill over from one country to another: what are the long-term environmental implications of Westernization, modernization and Americanization, and how are these tendencies modified and mediated by local traditions, habits, meanings and modes of appropriation? (Shove, 2003a, p. 9)

The above questions show that modernity in energy practice is conceptualised in two related ways. The first way is modernity as a discourse embedded in advertising, movies and other images in everyday life. People use discourses of modernity to give meaning to their practices and justify their behaviour. However, there is also another, more substantive dimension of modernity in energy practices, which is the continuous construction and re-construction of 'normality' in practices. This process, in which social practices change and become routinized and normalised through horizontal linkages, captures the essence of modernity in the energy practices literature, albeit in non-linear and contingent ways (Shove, 2003a, 2003b).

Assessment and Relevance of Energy Practices Literature for this Book

Energy practices literature conceptualises the dynamics of energy transitions by making 'practice' the key unit of analysis for understanding the changing demand for energy. Energy practice scholars understand energy transitions as the sum of these countless practices. They make a case for looking at what is going on in the construction of these practices in people's everyday lives, while recognising that practices extend well beyond individual behaviour. Indeed, as Watson (2012) shows, energy practices need not restrict analysis to the local or micro-level only, but can also be used to analyse issues 'across what can be understood as systemic scales' (p. 4). The implication for the study of energy transitions is to focus on how people actually use energy and how this shapes energy infrastructure, society and influences environmental sustainability. At the same time, the discourses of modernity are important to understanding how images of modern practices are formed through practices at different levels – such as ministries, company board rooms, by watching television – and how these in turn shape new socio-technical practices.

Energy practices literature has been framed as an alternative to the earlier discussed multi-level perspective for the understanding of sustainability transitions through many direct confrontations (see, for example, Geels, 2011; Rotmans and Kemp, 2008; Shove and Walker, 2007, 2010; Walker and Shove, 2007). These debates are helpful for gaining insight into the limitations of the socio-technical transitions literature, by exposing the non-linearity, contingencies and ambivalence of sustainable development and transitions, the failure of overly technocratic approaches to 'transition management', and the importance of horizontal processes such as standardisation and normalisation of practices. Moreover, rather than focusing on the role of actors and their power, energy practices literature emphasises that power is located in practices of standardisation and normalisation. For example, it is not the consumer who is 'in full control' of the temperature of his or her laundry, but also the companies producing washing powder and laundry machine, because they produce the manuals with 'appropriate' washing temperatures (Shove, 2004b).

Energy practices literature operates on a different scale and unit of analysis from the two other bodies of literature discussed so far (see also Table 2.1). The main emphasis of energy practices is on the consumption practices in and around the household, although it is certainly not confined to this scale and unit only. As Watson (2012) argues:

> While practice approaches have mostly so far found empirical application in relation to users and consumers and their ordinary doings, they equally have applicability to understanding the locales of action through which the rest of the systems of mobility are constituted. Practices recruit carriers in board rooms, the physical spaces of futures trading and government offices as much as they do on streets and in homes. (p. 496)

However, analyses of practices on these other locales have not as yet been the subject of much empirical research, making it difficult to assess whether the energy practices approach is able to live up to this claim.

To conclude, the conceptualisation of modernity in energy practice literature is very different from that in the socio-technical transitions and EMT literature. Rather than looking at loosely conceptualised changes at the 'landscape' level (as in the socio-technical transitions literature) or in 'a new social order' of reflexive modernity (as in Ecological Modernisation Theory), energy practice literature finds modernity in the changing relations between materials, competences and meanings, which together constitute practices and are constructed and reconstructed in everyday life. There are also some shortcomings vis-à-vis the energy practices literature. As shown in this chapter, few articles draw upon case studies outside of Europe and North America, despite the large potential. Moreover, notwithstanding recent contributions by Watson (2012) and others, the ability of energy practices to engage with processes on different 'scales' and units of analysis other than consumption practices is still underdeveloped. Political ecology literature, to which this chapter now turns, shares some of the concerns of everyday life and micro-level analyses. It also has a strong multi-scale framework and traditionally focuses on issues situated in the global South.

Political Ecology

Introduction and Conceptualisation of Transitions

Political ecology brings concepts of social justice, scale and emphasis on the role of capitalism[2] to the study of energy transitions, modernity and sustainability. One contemporary (loose) definition of political ecology is that it is an approach 'to address the condition and change of social/environmental systems, with explicit consideration of relations of power' (Robbins, 2012, p. 20). Perhaps the most diverse of all the approaches discussed in this chapter, it has gained widespread currency in disciplines such as anthropology, development studies, environmental politics, and human geography in particular over the last three or four decades. Therefore, it is difficult to provide a concise overview of the field, which some claim has deep roots in the form of environmental research carried out by nineteenth- and twentieth-century ecologists, ethnographers, explorers and other researchers (Robbins, 2012). However, political ecology emerged as a distinct academic field in the 1970s and 1980s.[3] The early development of the field is associated with the work of Watts (1983) and Blaikie and Brookfield (1987), which focuses on providing alternative explanations for environmental problems such as land degradation, soil erosion, deforestation, desertification and drought. Instead of following the then dominant explanations of overpopulation, mismanagement and market failure, these authors used a Marxist-inspired political economy approach focused on power, class and the state to analyse the above problems (Watts, 2009b). Blaikie and Brookfield famously defined their political ecology as 'the concerns of ecology and a broadly defined political economy' (Blaikie and Brookfield, 1987, p. 17), showcasing their commitment to understanding how ecological processes are influenced by economic and political systems and political ecology's strong materialist basis.

Twenty-five years later, political ecology has expanded in a large number of directions, using different definitions, methods and topics of research. Whereas early political ecology mainly focused on environmental problems in the 'Third World' (Bryant, 1998; Bryant and Bailey, 1997; Watts, 1983), political ecology has now been applied to many different issues and geographical locations (Neumann, 2005; Peet et al., 2011). A few key changes will be briefly highlighted here, the first being the influence of post-structuralism on political ecology, which

---

2 Capitalism is defined here as 'an economic system in which private capital or wealth is used in the production or distribution of goods and prices are determined mainly in a free market; the dominance of private owners of capital and of production for profit' (Oxford English Dictionary, 2012a).

3 The term 'political ecology' has also been used in other disciplines and related to other topics. For an overview, see Neumann (2005, pp. 3–6).

has shifted the approach to the study of discourses focusing on development and the environment. While discourse analysis is now well-established in many major publications on political ecology (Peet and Watts, 2004; Stott and Sullivan, 2000), its implications for epistemology and ontology are heavily debated and contested (Escobar, 1999, 2010; Watts, 2003). A related direction of expansion of the literature is towards the nature of scientific explanations and its links with Science Technology and Society (STS) studies, for example, through the work of Forsyth (2003, 2011). His 'critical political ecology' advocates more moderate forms of post-structuralism and post-modernism based on a critical realist perspective (cf. Sayer, 2000b). Another important development is the work on urban political ecology, which differs somewhat from the other forms of political ecology due to having its roots in critical urban studies and its focus on the 'developed world' (Heynen et al., 2006; Keil, 2003). The urban political ecology of Graham and Marvin (1994, 2001) is particularly relevant for this book as it provides a framework for analysis of urban infrastructure (such as electricity) from a political ecology perspective.

Much of the work undertaken on political ecology related to energy has focused on the interconnection between global capitalism and fossil fuels, particularly oil (Bridge, 2011; Pred and Watts, 1992). At the same time, there has been a trend towards a more holistic treatment of energy in the disciplines of human geography and political ecology. One example is the special issue on 'New Geographies of Energy' in the *Annals of the Association of American Geographers* journal (2011, volume 101, issue 4) which 'examines changing energy landscapes by combining the perspective of globalization processes operating at multiple scales – including many national and sub-national – with a focus on environmental change and resource systems' (Zimmerer, 2011, p. 705). Another example of energy-related research in political ecology is an article by Bradshaw (2010), who argues that human geography can make a significant contribution to solving 'global energy dilemmas' such as the transition to a low-carbon economy.

Many authors associated with political ecology have carried out research in Southeast Asia, albeit mostly related to the development of large hydropower dams.[4] Philip Hirsch, for example, was one of the first to publish academically on the resurgence of dam building in the region (Hirsch, 1988, 1996, 1998, 2010). While not necessarily employing the term 'political ecology', Hirsch's work carries some of its hallmarks: it is thoroughly multi-scalar – from ethnographic fieldwork on affected communities to analyses of national politics, regional integration and multilateral institutions – and underpinned by a social justice framework. Other scholars have contributed to the research into hydropower in

---

4 If one included 'grey' literature, the amount of political ecology work would much larger.

Southeast Asia using a similar approach, for example by focusing the discursive framing of hydropower on the Mekong river (Bakker, 1999); ecological understandings of river basins in the context of changing transboundary institutional arrangements (Sneddon and Fox, 2006); dam building, energy conservation; and civil society (Foran, 2007); hydropower as part of a contested waterscape (see several chapters in Molle et al., 2009b); and hydropower and the Clean Development Mechanism in Vietnam (Smits and Middleton, 2014).

There is also an increasing body of literature placing energy issues and energy politics more centrally in the analysis, from a (broadly defined) political ecology perspective. Examples from Southeast Asia include Chris and Chuenchom Greacen's work on micro-hydropower in Thailand and its links to the country's political economy (Greacen, 2004; Greacen and Greacen, 2004). A similar approach was adopted by Smits and Bush (2010) to understand the marginalisation of pico-hydropower, a small-scale renewable energy technology, in Laos. Other publications discuss energy demands in the context of energy politics and institutional arrangements in the Greater Mekong Subregion (Greacen and Palettu, 2007; Merme et al., 2014; Yu, 2003). Finally, the special issue on *Actors, Interests and Forces Shaping the Energyscape of the Mekong Region* in the journal *Forum for Development Studies* (2012, volume 39, issue 2) conceptualised the Mekong region as an 'energyscape' in which energy and hydropower are among the main uniting, yet contested, elements in the region (Kaisti and Käkönen, 2012).

Engagement with Social Theory and Modernity

Political ecology proponents draw upon a variety of social theories, reflecting the diversity and multi-disciplinary nature of the literature. The conceptualisation of modernity, however, was initially based mainly upon a rather deterministic neo-Marxist framework that explained the spread of global capitalism and its perceived negative social and environmental impacts (Bryant, 1998). Later, political ecology turned to more complex models of human-environmental interaction, which involved more emphasis on the state and other actors in shaping these interactions (cf. Forsyth, 2003), including post-structuralist influences. That said, it may be argued that capitalism, and more recently neoliberalism (Castree, 2006; Harvey, 2005; Heynen et al., 2007), are often still at the core of the understanding of modernity for many political ecologists.

The book that speaks most directly to modernity in political ecology is *Reworking Modernity: Capitalisms and Symbolic Discontent* by Pred and Watts (1992). In this work, the authors note that 'it should be made clear that modernism is no more a simple mirroring of modernization than modernization is synonymous with capitalist logic' (Pred and Watts, 1992, p. 13). This quote shows the connection that some key political ecologists make between capitalism and modernity and vice versa, or 'capitalist modernity' as Pred and Watts call it. This book is one of the early examples showing the influence of post-structuralism or post-modernism on political ecology. The authors talk about 'multiple modernities' and 'multiple capitalisms' to illustrate the way capitalism/modernity is embedded in different places and contexts, without a single origin. To put capitalism on the same level or even equate it with modernity requires a broad definition of the former, for example, including the realms of culture, politics and society. In a later definition of capitalism, Watts (2009a) notes that '[s]ome cling to a narrow definition of economy (theorised in different ways) as central to the intellectual enterprise; others seek to link economy, culture, politics and society into a more elaborated sense – what Max Weber called a "cosmos" – of a capitalist system' (p. 60).

David Harvey (1989), arguably the most prominent contemporary Marxist geographer, also links modernity with capitalism in his *The Condition of Postmodernity*, in which he frames the shift from modernity to post-modernity as a logical outcome of capitalism and capitalist accumulation. His framing of capitalism goes hand-in-hand with the dominance of neo-Marxist thinking which is still key to the work of many key geographers and (urban) political ecologists (Heynen et al., 2006; Mann, 2009; Swyngedouw, 2000; Swyngedouw and Heynen, 2003).

The influence of post-structuralism on political ecology did not necessarily alter this conflation of modernity and capitalism. This can be illustrated by looking at the work of Arturo Escobar (1996, 1999), one of the most influential scholars advocating for post-structuralist political ecology. Escobar's version of political ecology builds on an understanding of modernity as capitalism; for example, when he suggest that '[t]he history of modernity and the history of capitalism must be seen as the progressive capitalization of production conditions' (Escobar, 1996, p. 333). However, his approach focuses more on the role of discourse and the production of scientific knowledge, which underpins this coupling of capitalism and modernity and displaces other types of knowledge in the development of the global South: '[O]ne of the defining features of modernity is the increasing appropriation of "traditional" or pre-modern cultural contents by scientific knowledges, and the subsequent subjection of vast areas of life to regulation by administrative apparatuses based on expert knowledge' (Escobar, 1996, p. 333). Escobar, in essence, focuses on the Eurocentric aspects of capitalist modernity, disguised as the development industry, and the adverse impacts of it such as increasing inequality and the 'underdevelopment' of the global South (Escobar, 1995).

However, there are some serious problems surrounding the conflation of modernity with capitalism in political ecology literature. A good overview of the problems of this approach may be found in a chapter in the *Handbook of Cultural Geography*, in which Watts (2003) discusses the ways in which modernity has been used in cultural geography and the post-development movement. The key points of critique by Watts are the understanding of development/modernity as totalising power, the uncritical celebration of grassroots and new social movements, and the rejection of neoclassical economics without providing any alternative ways of thinking about economics. He concludes that the post-development movement 'has curiously not engaged sufficiently with the idea of development as modernity … [because] [m]odernity contains the tragedy of underdevelopment: development and its alternatives are dialectically organized oppositions within the history of modernity' (p. 441). This argument nuances some of Watt's own earlier work which perhaps focused too much on capitalism as the key driver of modernity (Woodward et al., 2009). Other examples of political ecologists who do not see capitalism as the only driver of change include Bryant and Bailey (1997), who maintain that '[t]he work of political ecologists has been largely an attempt to describe the spatial and temporal impact of capitalism on Third World peoples and environments … [y]et the source of the Third World's environmental woes cannot be equated with the workings of the global capitalist system alone' (p. 3).

Tim Forsyth's arguments in *Critical Political Ecology* (2003) are even more explicit against conceptualising capitalism alone as a driver of modernity and as explanation for environmental problems. He argues that the neo-Marxist critique of capitalism – and the way it has become part of social movements, NGOs and other institutions – has shaped debate about environmental degradation, making it part of environmental politics instead of 'objective analysis'. In other words, '[b]y questioning the essentialist link between capitalism and environmental degradation, [critical political ecology] challenges virtually all historic approaches to political ecology that have focused on political economy and environment' (p. 7). The last two quotes show the unease that some political ecologists experience when treating capitalism as the only explanatory factor and demonstrates a move towards more nuanced understandings of the dialectical relation of modernity and development. It also opens up more possibilities of thinking outside or beyond capitalism or neoliberalism (cf. Gibson-Graham, 1996, 2008; Thrift, 2005).

Assessment and Relevance of Political Ecology Literature for this Book

Political ecology literature provides many insights for this book, such as the emphasis of multi-scalar research as well as its engagement with social and environmental justice in the global South. Moreover, as pointed out earlier, there is strong emphasis on local fieldwork and agency, which partly resonate with the energy practices literature. The (partial) conflation of modernity as capitalism shows similarities with the Ecological Modernisation Theory literature, albeit as a negative relation and equally problematic. Some authors have proposed more nuanced approaches that make space for additional critical understanding of the relationship between modernity and capitalism, such as is apparent in more recent work in political ecology (Adger et al., 2001; Forsyth, 2003; Neumann, 2005; Zimmerer and Bassett, 2003). This would also open up political ecology to an understanding of sustainability which goes beyond seeing sustainability as a discourse which perpetuates existing power relations.

The review of the literature has shown that there are many potential points of overlap and complementarity between political ecology and the three bodies of literature discussed in this chapter. Yet, to date, few scholars have explored these overlaps in detail, perhaps due to the different geographical focal points between political ecology and the other three approaches. Among the exceptions are Watts (2000), who comments on the complementary nature of ecological modernisation and political ecology,[5] Lawhon and Murphy's (2011) recent article about the socio-technical transition approach and the insights of political ecology, and Walker and Shove's (2007) references to urban political ecology. In the following section, I will build on the key concepts of this book – energy, modernity and sustainability – and these modest points of overlap to come up with an approach for a critical scalar analysis of energy transitions in Southeast Asia.

Towards a Scalar Analysis of Energy, Modernity and Sustainability

The final part of this chapter refocuses the discussion on the key concepts introduced at the start of this chapter – energy transitions, modernity and sustainability – incorporating the issues discussed through the four bodies of literature.

Debating Modernity and the Energy-Modernity Dialectic

The discussions throughout this chapter have shown various interpretations and conceptualisations of modernity. Despite the sometimes contradictory uses in the four bodies of literature, there are some key ideas that this book draws upon from these discussions. First, modernity cannot be seen as a unified and coherent process. This is not only argued by advocates of multiple modernities, but also emerges out of the practice theory and some (critical) political ecology literature. Even when equated with the idea of capitalism, such as in Ecological Modernisation Theory and some strands of the political ecology literature,

---

5   Watts's understanding of ecological modernisation, however, is probably closer to the social-constructivist approach put forward by Hajer (1995).

the effects and implications are always uneven. This also serves to remind one that modernity does not necessarily lead to positive outcomes and should be conceptualised as a double-edged sword. These points resonate with Latour's (2003) suggestion that '[the] modern has lost two of its features: it no longer means something is incontrovertibly good and, more importantly, it is no longer associated with a coherent set of values and objects' (p. 42).

Second, an analysis of modernity is not something that can be reduced to one particular scale. The socio-technical transitions literature, for example, tries to include modernity in the 'landscape' level of its multi-level framework. However, as this chapter has shown, placing modernity outside of the direct scope of analysis weakens and de-politicises modernity. The other three bodies of literature put processes of modernity more central in their analyses, albeit in different ways. For EMT, it is assumed that reflexive modernisation has led to a 'new social order', which will eventually result in more sustainable development as environmental and ecological considerations become increasingly internalised within capitalist modes of production. For political ecology scholars, modernity is often conflated with capitalism and their research tends to focus on the negative impacts of capitalism on livelihoods, ecosystems and society. A more nuanced notion of modernity is found in the energy practices literature, which locates the idea of modernity in the continuous construction and re-construction of normality, which is the outcome of the everyday configuration of materials, competences and meanings.

Third, there is persistent Western or Eurocentrism in the conceptualisation of modernity in some of the bodies of literature used in this chapter. For example, Beck – when discussing the notion of reflexive modernity – persistently argues that modernity has European origins and that reflexive modernity of necessity originated in the West/Europe. However, others have argued that the use of modernity in this way implicitly marginalises developments outside of Europe and creates a single global narrative (Massey, 2005; Strohmayer, 2009). This book retains the concept of modernity, following the argument proposed by Watts (2003) that modernity not only can be found anywhere in the world: it even has distinct origins in various parts of the world (cf. Chakrabarty, 2000). This argument can also be found in the ideas of alternative modernity and multiple modernities.

This chapter argues for an understanding of energy transitions as intertwined with modernity, or, as the energy-modernity dialectic. As suggested at the start of this chapter, Harvey (1996) describes a dialectical relation as two elements that are continuously in flux, where change in one affects change in the other, and neither takes primacy. Drawing upon the four bodies of literature discussed, one can identify several arguments to support this dialectic. First, the concept of socio-technical systems and the seamless web in the work of Hughes (1983, 1986) shows the inseparability of technical artefacts and social relations. This means that technology shapes social relations, society and vice versa, a proposition captured, for example, in the book title *Shaping Technology / Building Society* (Bijker and Law, 1992). The idea of such a dialectical relation can also be found in actor-network theory (ANT), in which actor-networks are conceptualised as webs of relations between human and non-human elements and where social change occurs through the translation, enrolment, ordering and re-ordering of these elements (Law, 1992; Law and Hassard, 1999).

Modernity and energy are also closely linked in the political ecology and Ecological Modernisation Theory literature, which both conceptualise modernity as having important connections to the production and consumption of energy and other material aspects. Energy is one of the key aspects of environmental reform in EMT studies, which in turn are seen as one of the outcomes of reflexive modernity. Political ecology, on the other hand, focuses more on local impacts of large-scale energy production in the context of capitalist modernity, as well as on discursive aspects of human-environment interactions and the role and politics of discourse and knowledge.

The Place of Sustainability in the Energy-Modernity Dialectic

Each of the four bodies of literature demonstrates different ways to understand how social and technical developments are interrelated and how they ultimately affect environmental sustainability. For transition management scholars – who argue for deliberate steering towards sustainability transitions – sustainability is often a normative and rational goal that may be reached by fostering technologies in niches to overcome the lock-in effects of the extant socio-technical regimes (Scrase and Smith, 2009; Shove and Walker, 2007; Smith and Stirling, 2010). In doing so, it places environmental sustainability outside of the socio-technical system or energy-modernity dialectic. However, this may prove problematic because it turns sustainability transitions into an apolitical process which is 'rendered technical' and subject to technological or bureaucratic fixes (Li, 2007a). Political ecology scholars who conflate modernity with capitalism, as reminiscent of its rather dogmatic neo-Marxist origins (Bryant, 1998), are in danger of putting sustainability outside of the energy-modernity dialectic, albeit in a different way. They see sustainability and sustainable development as a discourse which serves to mask 'real' political power/knowledge, rather than something with a discursive and material basis.

In Ecological Modernisation Theory, energy practices, and some political ecology literature, sustainability is more internal to energy-modernity dialectic. EMT employs the powerful notion of reflexive modernity, but the focus of this literature often leans heavily towards the positive examples of ecological modernisation rather than viewing modernity as a double-edged sword with sometimes positive but also negative sustainability implications. Energy practices literature provides another way of integrating sustainability into its conceptual framework through its appreciation of the construction of normality in everyday practices which involves materials, competences and meanings. This is useful insofar as it considers the convergence of certain established global practices, but less so for different contexts and multi-scalar processes. Political ecology literature could provide a framework for putting sustainability at the heart of the energy-modernity dialectic, provided that it avoids the pitfalls of extreme post-modernism on the one hand, and structural Marxist interpretations on the other.

Instances of Modernity

This chapter has reviewed the key concepts of this book – energy transitions, modernity and sustainability – and looked at how each of them has been conceptualised in the four bodies of social science literature. The review shows that an understanding of modernity and its different conceptualisations are important to understanding how energy transitions shape society and vice versa. There are important differences within and between these bodies of literature, for example in how they conceptualise energy transitions, modernity and sustainability. A closer look reveals that some of the differences may be overcome by looking at the underlying concepts, assumptions and social theory. In this book, modernity is understood as an ongoing process without fixed origin or end point, a process dialectically related to energy. The idea of reflexive modernity provides a potential avenue to integrate sustainability into this dialectic, albeit not in the way of a rupture in as Beck et al. (2003) would have it. Rather, modernity is enacted at different and interrelated scales and in discourses, with capitalism playing a key – but by no means all-encompassing – role.

The discussions about modernity feature throughout the rest of the chapters on different scales and across different dimensions. Table 2.2 summarises these different 'instances' of modernity for the country-level and local-level cases. Then, in Chapter 3, discourses of modernity and sustainability in Southeast Asia, Thailand and Laos are analysed to discern how they have shaped society and energy sector development, using concepts such as state-formation, territorialisation and the geography of cost and benefit. This chapter not only focuses on the discourses themselves, but also on how they become embedded in material dimensions of modernity, such as energy infrastructure. Furthermore, it looks at alternative discourses of energy-modernity and how they lead to renewable energy policies and projects. In Chapters 4 and 5, a micro-level approach to modernity is taken, inspired by the energy practice literature. In these chapters, I analyse the four case studies as energy trajectories, by discussing the changing practices and increasing invisibility of energy infrastructure, but also explore experiential dimensions of modernity such as worldviews, culture and perceptions of time and distance. Finally, in Chapter 6, modernity again takes centre stage to see how (reflexive) modernity shapes and co-constitutes energy transitions, energy-modernity and sustainability, at and across different scales in Southeast Asia.

Table 2.2   Instances of modernity featured throughout the book

Dimensions	Regional/country-level cases (Chapter 3)	Local-level cases (Chapters 4 and 5)
Spatio-political	State-formation and territorialisation Changing geography of cost and benefit	Integration in state and markets
Discursive	State-led development discourses and its alternatives Discourses of sustainability and regional cooperation	Changing worldviews
Material	Diffusion of infrastructure, appliances, media and other technology	Changing energy practices
		Increasing invisibility of energy infrastructure
Experiential		Changing worldviews and influence on cultures
		Changing experiences of distance and time
		Individualisation

Source: Author.

Chapter 3

Regional and National Energy Transitions in Southeast Asia

Introduction and Key Arguments

This chapter serves the dual purpose of providing the background to developments in the energy and power sector in Southeast Asia from the mid-nineteenth century – with a specific focus on Thailand and Laos – and analysing the discourses of energy-modernity and sustainability in this region. My reason for going back to the nineteenth century is to trace the history of the power sector back to its first usage, its institutionalisation, and resultant modes of governance in this region. Modernity in this chapter is mainly understood as a discourse that has been used by various actors to justify their decision-making regarding development in general and in the power sector in particular. However, this chapter also shows that modernity is a double-edged sword, which while changing social relations, worldviews and practices, also re-confirms old hierarchies, entrenches vested interests and, in the process, limits freedom.

The first key argument of this chapter is that energy transitions in Southeast Asia are increasingly shaped by regional discourses of energy-modernity, such as through ASEAN and ADB's Greater Mekong Subregion projects. In part, this increasing importance is due to fact that energy demand keeps increasing and there is generally more trade and cooperation between nation states in Southeast Asia. This is epitomised in the various ASEAN energy plans and activities, such as the regional power grid. Beyond the nation state, the ADB is playing an important part to develop a shared regional identity in the GMS and promote the development of infrastructure through private investment. Energy infrastructure – large-scale hydropower plants and high-voltage transmission lines in particular – is a key component of the GMS project.

Second, and zooming in, transitions in the electricity sector in Thailand and Laos are closely related to the conflicting discourses and materialisations of 'modernity', following the context-specific processes of state formation and territorialisation in these two countries. While Thailand itself was never colonised, the period of colonisation in Asia in general led both countries to establish borders and to the inclusion of people into the nation state since the nineteenth century (amongst other means) through infrastructural developments such as (rural) electrification. State-led discourses of modernity, which were dominant in both countries until the last half of the twentieth century, have increasingly been challenged by civil society and environmental movements, in particular in Thailand. The counter-discourses focus on decentralised and local forms of (renewable) energy production. In Laos, however, such arguments are much weaker and mainly come from international NGOs and bilateral donors. However, changes are slow due to the long construction periods and high capital investment of infrastructure, entrenching power relations and patterns of inequality for decades or longer.

The third key argument is that energy-modernity is not only shaped by both deliberate and top-down processes such as state formation, territorialisation and regional integration, but also by people and their everyday practices, some of which are unintended consequences of the above policies. In other words, one needs to look at how energy systems shape society and vice versa. This chapter does so by looking at the ways in which people have used different forms of energy and how they have affected their energy practices such as lighting, (mass) media and transport. The resultant increased levels of education, decreasing distances, and new forms of communication have challenged and reinforced the ways in which power is exercised and, in some cases, has forced or facilitated the creation of new forms of governance. These points, as well as undermining the assumption that energy transitions unfold in a linear, progressive or top-down fashion, stress their contingent and non-linear nature.

The rest of the chapter is structured as follows: it starts with section on discourses of regional energy-modernity and integration in Southeast Asia, focusing on the role of ASEAN and the GMS project. Then, the chapter zooms in on energy and modernity in the power sector in Thailand followed by a section on Laos.

History and Regional Discourses of Energy-Modernity in Southeast Asia

This section discusses the history and discourses of regional energy-modernity and power trade, focusing on ASEAN and the Greater Mekong Subregion (GMS). These regional discourses open up space for the territorialisation of a specific energy-modernity mainly through Independent Power Producers (IPP) and centralised interconnected electricity grids.

History of Bilateral Electricity Trade in Southeast Asia

While electricity was introduced to many Southeast Asia countries at the end of the nineteenth century, cross-border trading of electricity was very limited in Southeast Asia until the late twentieth century. Changes started to occur from the 1960s on for a number of interrelated reasons, the first one being the increase in political stability over the previous two decades and increasing political cooperation. After the Second World War, the countries in Southeast Asia were divided according to the dictates of the Cold War superpowers and by the wars in Indochina. At the time, many of the governments were inward-looking and not open to regional cooperation, or only cooperated between similarly aligned states. This changed after the conflicts were resolved and regional platforms started to include countries previously aligned with either the 'capitalist' or 'socialist' worlds. This change was captured in an iconic comment by former Thai prime minister Chatchai in 1988 when he spoke of 'turning battlefields into marketplaces' (Yu, 2003).

A second and related reason was the increasing pace of development and energy demand in the region. As Jerndal and Rigg (2000) argue, '[t]he process of modernisation, it seems, is regarded as an integrating force in the area' (p. 39). Thailand was among the countries where rapid industrialisation started from the 1960s on, whereas countries such as Laos would follow this trend later, resulting in a strong increase in energy demand across the whole region. Concomitant with this increase in energy demand, the need for new sources of fossil fuels, for example oil and gas, as well as hydropower, increased. Figure 3.1, which shows the magnitude of these developments for the electricity sector of Southeast Asia, demonstrates some notable energy transitions, such as the introduction of natural gas in the 1980s that accounted for almost half of all the electricity production in 2008. The share of coal-based electricity generation also increased rapidly during the same period, while the shares of hydropower and electricity from renewable and waste sources were relatively small. The final reason for

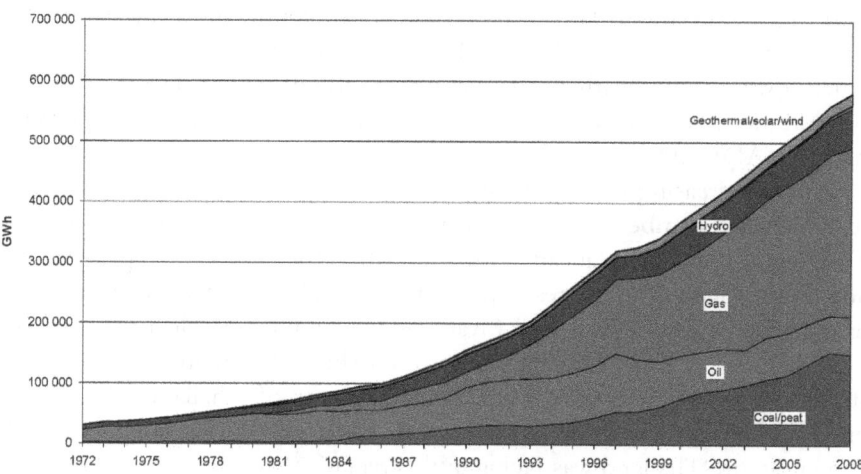

Figure 3.1　Electricity generation by fuel for Southeast Asia/ASEAN

Source: IEA Energy Statistics, www.iea.org/statistics © OECD/IEA [2010], IEA Publishing.

cooperation is that, while the region is rich in energy resources, 'these resources are located in areas that are far from the major markets of electricity demand and separated by national borders' (Yu, 2003, p. 1,222).

Third, neoliberal discourses, such as deregulation, privatisation and foreign investment, proved important to increasing the speed of regional integration of the power sector (Williams and Dubash, 2004). During the Cold War, both superpowers, as well as the multilateral institutions, supported the establishment of state-owned utilities, or vertically-integrated monopolies controlling generation, transmission, distribution and retail. This changed after the rise of neoliberalism and the oil crises. Neoliberalism in the electricity sector was first trialled in the 1980s by Chile and the UK. It was subsequently picked up by the Bretton Woods institutions and linked with their Structural Adjustment Policies, loans and other policy measures under the Washington Consensus (Sheppard and Leitner, 2010). As Williams and Dubash (2004) argue:

[t]he transition from a state-led development model to one emphasising private ownership and competitive markets was less a targeted response to specific conditions in any Asian country's power sector than a reflection of the Washington Consensus belief that the private sector should be primary and the state secondary in all areas of the economy. (p. 433)

While pressure for more reforms increased after the Asian Financial Crisis in 1997, the California Electricity Crisis in the early 2000s led to the easing of the most radical reforms (Williams and Dubash, 2004). However, the partial reforms that were adopted – for example the introduction of IPPs, the partial privatisation of utilities, and the opening up of the economy to foreign investment – were important drivers for regional cooperation and integration. By 2000, the ADB acknowledged that '[t]here is growing interest in cross-border power trade, spearheaded by regional and international developers who have commenced hydropower development in the Lao PDR to sell power to Thailand' (ADB, 2000, p. 2).

Fourth, increasing environmental awareness, the strength of civil society, and increasing numbers of protest movements have altered the eco-politics in the region, providing an additional incentive for regional cooperation and trade in electricity. The domestic pressure in Thailand underpinned its search for energy projects in Laos and Myanmar for supplies of electricity and gas. While this reason is not often articulated in official documents, my interviews with people in the Electricity Generating Authority of Thailand (EGAT), the Ministry of Energy, and in the province of Prachuab Khiri Khan, revealed the strong pressure Thailand was facing in this regard.

Initially, cooperation between the countries of Southeast Asia was mainly through bilateral deals, especially low-voltage cross-border distribution lines.

As noted above, in 1966, Thailand and Laos signed the first high-voltage bilateral electricity agreement in the region. Further bilateral electricity trading agreements were completed between Thailand and Malaysia and Malaysia and Singapore in 1978 (Nicolas, 2009). The next sections details how more of such deals became part of the regional discourses of ASEAN and the GMS programme from the 1980s onwards.

The Discourse of Regional Electricity Cooperation by ASEAN

The Association of Southeast Asian Nations (ASEAN) has played an increasingly important role in promoting discourses pertinent to regional power trade. When ASEAN was founded in 1967, it included Indonesia, Malaysia, the Philippines, Singapore and Thailand, that is, only five of the current ten member countries. The objective was to 'promote greater economic cooperation in the region and also present an organised bloc against the threat of communism' (Sovacool, 2009b, p. 615). Among the early collective activities in the energy sector of this US-aligned organisation was the setting up of a petroleum council in the wake of the oil crises of the 1970s. The first ASEAN energy ministers meeting was held in Bali in 1980 (ASEAN, 1980). One year later, the Heads of ASEAN Public Utilities forum was set up, bringing together the different utilities, with the objective to establish further regional cooperation and investigate power grid connections (Nicolas, 2009).

Following the signing of the ASEAN Energy Cooperation Agreement in 1986, cooperation between the ASEAN countries became more diverse and intense. Around this time, Laos and Vietnam started to open up their economies and the Cold War era came to an end. As a result, ASEAN expanded to include Brunei (1984), Vietnam (1995), Laos (1997), Myanmar (1997) and Cambodia (1999). After the Asian financial crisis of 1997, ASEAN energy ministers initiated meetings with China, Japan and South Korea to coordinate energy issues in the ASEAN+3 meetings. In 1999, the ASEAN Centre for Energy was set up, and cooperation extended to include other regional powers (Nicolas, 2009).

According to Sovacool (2009a), the energy vision of ASEAN is to 'create a regional, harmonised framework of energy supply that utilises commodities of coal, oil, natural gas, and electricity to promote industrialisation and economic growth' (p. 2,357). Since the late 1990s, trade in electricity and gas has been the main priority under the regionalisation discourse, operationalised in an 'ASEAN Plan of Action for Energy Cooperation' for every five years. This section discusses the three first plans, covering the periods 1999–2004, 2004–2009 and 2010–2015 (ASEAN, 1999, 2004, 2010). The fields of cooperation have been more or less the same in all three plans: (1) ASEAN Power Grid; (2) Trans-ASEAN Gas Pipeline; (3) Coal and Clean Coal Technology; (4) Energy

Efficiency and Conservation; (5) Renewable Energy; (6) Regional Energy Policy and Planning. Only one, item (7) Civilian Nuclear Energy, is new in the last plan, although some fields have been rephrased (for example from 'Coal' to 'Coal and Clean Coal'). The order of the fields of cooperation provides an indication of the order of priority of ASEAN regional cooperation, namely emphasis on infrastructural interconnections through the ASEAN Power Grid and the Trans-ASEAN Gas Pipelines. These issues also feature in general ASEAN meetings and plans, for example 'ASEAN Vision 2020' from 1997, the Hanoi Plan of Action from 1998, and the Vientiane Action Plan from 2004.

Despite all of these plans and the rhetoric about creating a common market, there are still many obstacles to overcome, alluded to by policy makers in Thailand during interviews. One of the Commissioners of the Energy Regulatory Commission in Thailand, for example, commented that 'in Cambodia, there is no electricity in the countryside. In Myanmar, they still use candles. How can we have a market in the region?' Some commentators remain sceptical about these regional energy modernity discourses due to the current fragmented electricity markets and the persistent focus on bilateral agreements. Sovacool (2009a) argues to the effect that 'talking about regional energy cooperation is much easier than actually cultivating it' (p. 2,365). Moreover, Nicolas (2009) maintains that: '[a] number of signs tend to suggest that some countries in the region are not ready to go beyond paying lip service to regional cooperation because national interests still prevail over regional objectives' (p. 28). Among the national interests inhibiting increased regional cooperation to which Nicolas (2009) refers are the discourses of energy security, competition and rivalry between ASEAN countries over resources and other strategic interests (for example their relations with China). Nevertheless, ASEAN constitutes as an increasingly important scale to understand energy and modernity in Southeast Asia. The 10 members states are moving towards the establishment of an 'ASEAN economic community' by 2015; their ambition being to become 'a competitive single market and production base' and 'a region of equitable economic development fully integrated into the global economy' (ASEAN, 2008, p. 6). The next paragraphs look specifically at another scale, the Greater Mekong Subregion (GMS).

ADB's Discourse of the GMS Power Grid

The Greater Mekong Subregion – consisting of Cambodia, Laos, Myanmar, Thailand, Vietnam, and the Yunnan and Guangxi provinces of China – is another important scale at which cross-border electricity trade has been discussed and planned since 1992. As a discourse, the GMS shows parallels with the discourse of ASEAN, supporting the territorialisation of the region through neoliberal arrangements that favour large-scale power stations and integrated electricity grids. However, more so than the ASEAN power grid, the plans for infrastructural interconnections and power trading of the GMS are initiated and driven by external actors, chiefly the Asian Development Bank (ADB, 2011) and the World Bank (World Bank, 2007), but also supported by regional energy utilities. Like ASEAN, the regional discourse of the ADB focuses strongly on infrastructure and economic development, such as road and rail construction, trade, investment, water and labour (Jerndal and Rigg, 2000; Yu, 2003).

The main strategies that the ADB and the World Bank employ to materialise their regional and neoliberal discourses include the facilitation of regional platforms, knowledge production and sharing, and strategic loans (Glassman, 2010; Oehlers, 2006). This signals a marked change with the past, when these banks were often the main providers of funding for large scale infrastructure projects such as hydropower. According to Middleton et al. (2009), two main forces underpin this change. First, the roles of these 'old actors' are increasingly scrutinised and criticised by NGOs, media and by their constituting governments. Second, many 'new actors' have entered the financing scene, some of whom are willing and able to fund large scale hydropower and other projects, such as IPPs. These new actors are often subject to less stringent social and environmental safeguards and scrutiny compared to the multilateral Banks (Middleton et al., 2009).

Cross-border trading of electricity has been one of the key aspects of the GMS discourse since 1992, and the ADB sees it as one of its 'flagship projects' (ADB, 2005). One of the first activities implemented under the GMS programme was the undertaking of a sub-regional energy sector study. The year 1995 saw the commencement of the Subregional Electric Power Forum meetings. An integrated transmission system study, which focused on the Lower Mekong Countries and was completed in 1996 by the Mekong River Commission (MRC), resulted in a power trade strategy produced by the World Bank in 1998, as well as the creation of the Experts Group on Power Interconnection and Trade in 1998 (ADB, 2000). However, the Asian Economic Crisis of 1997 significantly lowered the energy demand projections. Therefore, the Norwegian hydropower consultancy company Norconsult was selected to redo the study (at the cost of US$1.4 million). The results were subsequently endorsed by the GMS countries in 2002 and the Regional Power Trade Coordination Committee was established (Ryder, 2003). Following this, the GMS leaders signed a Memorandum of Understanding on a Regional Power Trade Operating Agreement in 2005, providing some ground rules for power trading (International Rivers, 2007). The plans included an analysis of the electricity flows between each of the countries (Figure 3.2). Central to these plans were large numbers of high-voltage transmission lines between the countries. In 2009, there were seven such points of interconnection in the GMS and seven more projects planned (ADB, 2010). The next major step in GMS power trade cooperation was the establishment of

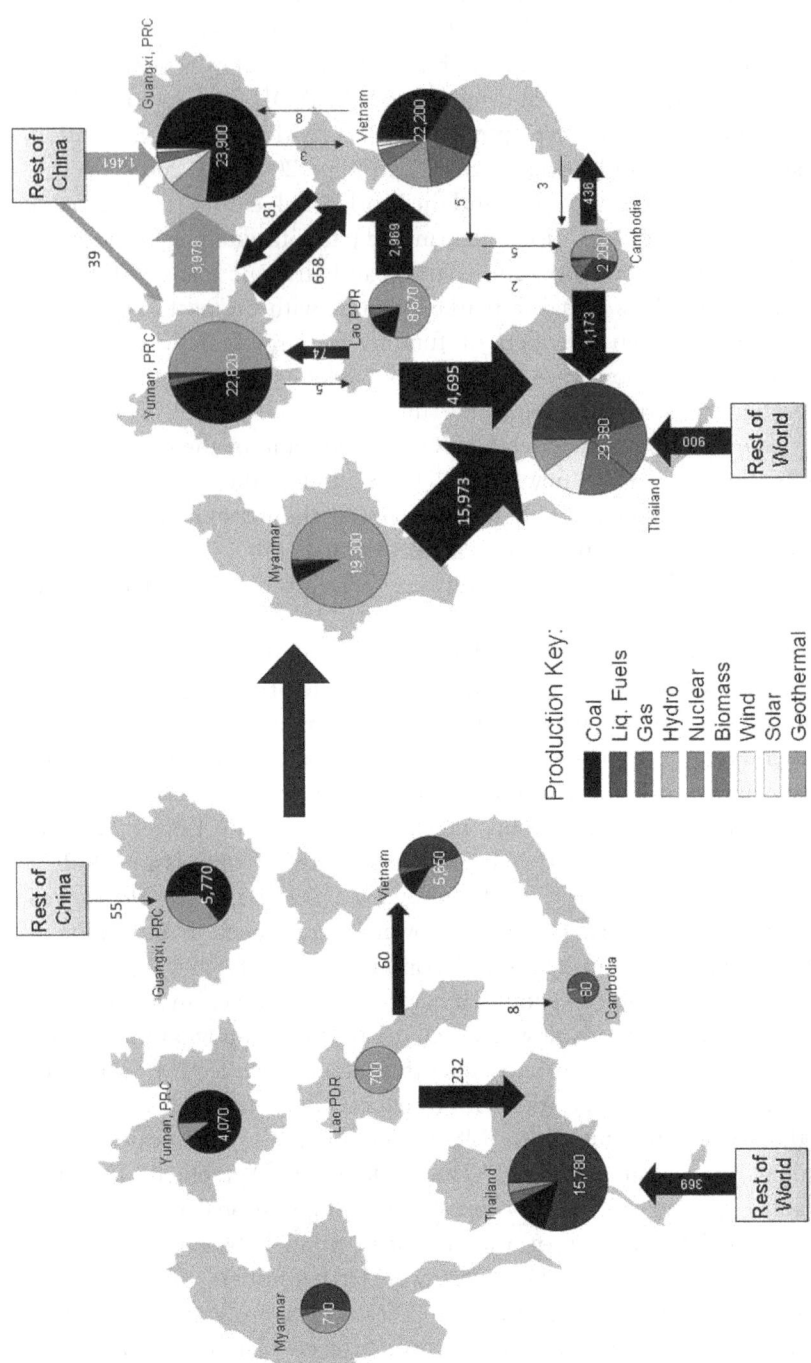

Figure 3.2 Electricity trade in the GMS in 2005 (left) and 2025 (right) when fully integrated. All units are in MW

Source: © 2009 Asian Development Bank. Adapted from ADB (2009). © ADB.

the Regional Power Coordination Centre Headquarters in 2015 (ADB, 2014). The regular meetings of the Committee and the establishment of the centre demonstrate how ADB's discourse of power trade and regional cooperation gradually becomes institutionalised and materialised.

According to the ADB, the interconnection of electricity networks at the sub-regional level would provide significant economic and environmental benefits for individual countries, as well as for the region as a whole. The main envisioned benefit was the reduction of national investment in power reserves to meet peak demand, in addition to: (1) increased reliability of supply in case of power failure; (2) lower operational costs; (3) less greenhouse gas emissions and pollution; and (4) access to 'the cheapest and most environmentally sustainable sources of electricity in the subregion, i.e., hydropower and gas' (ADB, 2000, p. 1).

This discourse of a neoliberal regional energy-modernity – and an integrated power grid based on centralised large-scale hydropower – has been challenged by NGOs, civil society organisations, and academics. These actors highlight not only the negative economic, social and environmental impacts of the power plants themselves, but also the impacts of the hundreds of kilometres of high-voltage transmission lines this involves. A third point is the neoliberal assumption of full market-based competition in the region under an independent regulator. Critics claim that this is problematic and unrealistic in a region dominated by large utilities with de facto monopoly positions, such as EGAT and EdL. As well, they insist that the high inequality in terms of access to electricity in the GMS will have to be resolved before releasing market forces in the electricity sector. Finally, they point to the lack of public participation in the development of the GMS power grid and other parts of the GMS programme (Garrett, 2004; International Rivers, 2007; Ryder, 2003).

In order to understand how regional cooperation fits in and is shaped by the development of energy-modernity discourses on national level, I now turn to two specific case studies of Thailand and Laos. These chapters not only show the importance of national processes of state formation and territorialisation, but also of the role of local people and consumers in shaping the energy transitions in these countries, with many parallels throughout other Southeast Asian countries.

Modernity and Transitions in the Power Sector in Thailand

1850s–1910s: The Rise of the Modern Nation State and the Start of Electricity Generation

In the mid-nineteenth century, Siam (now Thailand) started its transition from a feudal kingdom to a 'modern' nation state (Baker and Phongpaichit, 2014). This period was also seen as the start of a 'technocratic tradition' which reformed many of the 'traditional' governance structures (Stifel, 1976, p. 1,184). Therefore, it may be seen as the first period in Thai history in which the state took the lead in disseminating the discourse of state-led modernity. It was also during this period that the first electricity systems were developed and, as such, became part of this transformational period. One of the key elements of this period was the attempt to extend the influence of the kingdom beyond the capital to include other areas and vassal kingdoms. During this period, the people of Thailand were increasingly integrated as citizens of Thailand, with all the rights and duties that came with citizenship. Electricity, along with roads, railways, telegraph and mapping were very important in the processes of modernisation, state-formation and territorialisation (Winichakul, 1994).

While the region now known as Thailand has a long history of settlement and culture, during the greater part of the nineteenth century most of the country was covered by forest, and the borders between the different governed areas (*muang*) and other kingdoms were fluid (Winichakul, 1994). The population density was very low: most people practiced subsistence farming, with the exception of those living in the urban centres. Trade had been in progress for many centuries, but the volume was low and the infrastructure connecting the different *muang* was limited. Absolute monarchs ruled 'through, with, and at times against as a semihereditary elite, each combining military, administrative, judicial, and financial roles ... [creating] an arbitrary, "self-sustaining" diffuse political authority' (Girling, 1981, p. 46).

The transformation of Thailand into a 'centralised, functionally differentiated bureaucratic state' (Girling, 1981, p. 46) started during the reign of King Mongkut (Rama IV, 1851–1868) and was further developed by his son King Chulalongkorn (Rama V, 1868–1910). Chulalongkorn's ideas about progress were heavily influenced by his visits to other parts of Southeast Asia and Europe and the knowledge that came from outside, for example from foreign advisers (Stifel, 1976).[1] Baker and Phongpaichit (2014) argue that '[o]ver his 42-year reign the old political order was replaced by the model of the nation-state' (p. 52). In effect, Thailand had little choice but to engage with the main colonial powers, the British and the French, who were rapidly advancing on the region. In this sense, the signing of the Bowring Treaty of 1855 'marked the emergence of modern Thailand and its enmeshment in trading patterns negotiated on Western terms and involving Western notions of free trade' (Hewison, 2006, p. 81).

One of the ways in which this transformation took place was through closer links between Bangkok and the provinces. To achieve this, Chulalongkorn

---

1 A good example here is the hiring of the Dutch engineer Homan van der Heide, who tried but failed to transform Siam's irrigation system (Ten Brummelhuis, 2005).

enlisted the help of Prince Damrong, who travelled through all of the provinces with the aim to replace the traditional governance systems based on local patronage by provincial and district governors, who functioned under the Ministry of Interior. Moreover, he formalised the idea of the 'household', which meant registering all births, deaths and marriages, took over the use of force previously controlled by the local elite, promoted education, and drew the connection between the Buddhist religion and the state (Vandergeest and Peluso, 1995; Wyatt, 1984). Chulalongkorn also abolished slavery in Thailand as part of this modernisation programme, which was in part due to international pressure and to the adoption of a Western legal system (Feeny, 1989). These kinds of developments 'all reinforced the idea that all inhabitants of Siam were subjects of a single king, members of a single body politic' (Wyatt, 1984, p. 217).

Modern technologies, such as roads, ports, railways and telegraph – and later electricity – were instrumental in accompanying the discourse of state-led modernity, which was institutionalised under King Chulalongkorn (Baker and Phongpaichit, 2014). In 1858, the first steam-powered rice mill was built in Bangkok, followed by a steam-powered sugar mill in 1862. A post and telegraph service was initiated in the 1880s, and railways were built from Bangkok to Korat in 1900, to Chiang Mai in 1921, and to Khon Kaen in 1933 (Baker and Phongpaichit, 2014, pp. 83–9). These infrastructural projects were very important for the successful governance of the Kingdom, for as Feeny (1982) notes:

> The existence of railroads helped the government in its effort to administer the country closely in order to prevent abuses and local problems from providing excuses for foreign intervention and colonisation. Railroads allowed the government to make better use of its limited military power, and enhanced Bangkok's control over provincial and local government ... The railway provided security and public administration benefits to the élite and the nation and served as an important symbol of modernity. (pp. 80–81)

This period of reform coincided with the introduction of electricity to Thailand and elsewhere in the world. As early as 1884, an army general acquired a generator and electrified some of the military barracks in Bangkok and later the houses of King Chulalongkorn and other members of the Royal Family. This was shortly after the introduction of commercial electricity generation in New York and it marked a shift from biomass (fuel wood) and lamp oil use to 'modern' sources of energy for lighting worldwide (Hausman et al., 2008). The next evidence of electricity was a project introduced by Danish company Siam Electric Company Limited to run an electric tram in the vicinity of the Thai Royal Palace (Greacen, 2004). Just before the turn of the century, Thailand got its first power plant at Wat Leab, operated by the same Danish company which made electricity available to a limited number of the Bangkok elite. This particular plant was followed by another in 1912, which supplied electricity to the northern part of Bangkok. While these early power plants mainly used fuel wood, they also used coal (from Mae Mo), oil and rice husk. For many years, the above mentioned installations remained the country's only two power plants. Development slowed down during and after the First World War, between 1914 and 1918 (Sukkumnoed, 2007, p. 115).

These energy transitions were not only related to state formation and territorialisation, but also drove the first developments of mass culture, which would later become an important means to disseminate discourses of modernity. In 1897, for example, the first commercial film was screened in Bangkok, just two years after the first one screened in Paris. Film screening had become a regular and popular activity by the 1910s (Baker and Phongpaichit, 2014, p. 106).

1910s–1950s: Political Modernisation and Expansion of Electricity Generation

In the first half of the twentieth century, some major unforeseen political consequences of the institutionalisation of the modern nation state eventuated, in the midst of the two World Wars and the Great Depression of the early 1930s. In Thailand, while the style of governance remained centralised, absolute power was eventually taken from the King after a coup d'état in 1932. Moreover, the rise of the army as power broker, and the calls for more democratic participation during this period would prove to be important factors in shaping new discourses of state-led modernity in the power sector in Thailand. During this time, territorialisation continued through the electrification of the first provincial centres: infrastructure in general became increasingly important to sustaining the nation state and new industrial developments.

While King Chulalongkorn's successors tried to further modernise Siam under the absolute monarchy, this model eventually failed. One possible cause was that the two successors of Chulalongkorn, King Vajiravudh (Rama VI, 1910–1925) and King Prajadhipok (Rama VII, 1925–1935) were less successful as absolute monarchs and leaders of the country. For example, they failed to generate sufficient investment in industry and lacked interest in irrigation, transportation and education (Nartsupha et al., 1978). Moreover, the Great Depression of 1929–1930 led to a large reduction of the farmers' incomes due to the low price of rice. Under these circumstances, a small group of Western-educated men, supported by factions of the army, came together and overthrew the absolute monarchy in 1932. However, rather than a seeing this as a revolution, Baker and Phongpaichit (2014) claim that this coup 'created a modified and more powerful version of the strong-state tradition minus only the monarchy' (p. 138). A period of political instability followed, which ended with a coup staged by the Field Marshal Phibunsongkhram (abbreviation: Phibun) in 1938 (Baker and Phongpaichit, 2014; Wyatt, 1984).

Despite the political unrest caused by the coup, there were some noteworthy developments in the power sector during this period, namely the expansion of electricity into areas outside of Bangkok. The first provincial centre to get its power station was Ratchaburi in 1927, followed by Nakhon Phanom in 1930 and Chiang Mai in 1931 (Wattana et al., 2007). In 1929, a separate Electricity Division was set up by the Ministry of Interior, which dealt with the expansion of electricity services outside of Bangkok (Sukkumnoed, 2007, p. 116). These developments went hand-in-hand with the start of local industrial manufacturing of mass consumer products such as 'matches, nails, soap, fireworks, patent medicines, tobacco, oil, textiles, paper, bricks, fireworks, shoes furniture, and services such as power, water, and transport' (Baker and Phongpaichit, 2014, p. 93). Hewison (2006) argues that '[t]he overthrow of the monarchy brought new economic and political ways. Although still not a fully-fledged capitalist system, the commercialisation, monetisation, and commodification of the economy were well established by the 1930s' (p. 83).

Besides mass consumer products, new print media were also developed, sometimes reinforcing, but also challenging, state-led discourses of modernity. Translations of Western novels and romances, which were already popular before the turn of the century, were complemented by various local newspapers and magazines during this period. By 1927, Thailand had 127 printing presses and 14 publishers by 1927, and the first Thai film was produced in the same year (Baker and Phongpaichit, 2014, pp. 106–7). To counter the influence of these new media and urban practices – many of which came from abroad – the state increasingly started to interfere with issues related to the nation and its culture, especially in the lead up to the Second World War. Nationalism was cultivated through renaming the country 'Thailand',[2] the spread of national symbols, the adoption of an anthem, and modern standards for dressing, social life and public conduct. However, these interventions were largely restricted to Bangkok and a few other urban centres (Baker and Phongpaichit, 2014, pp. 130–34).

During the Second World War, Thailand allowed Japanese troops to move through the country. This alliance also helped Thailand to regain some 'lost territories' in present-day Myanmar and Laos. As the war progressed, Japan turned the tables on Thailand and occupied the country, by extension putting a temporary halt to further material and cultural modernisation projects of the state (Girling, 1981). The country's economy deteriorated heavily and, as a result, rural electrification plans came to a halt (Greacen, 2004, p. 127). The

---

2   Field Marshall Phibun renamed the country Thailand in 1939 as part of his nationalist politics, but this was revoked for a brief period following the Second World War. In 1949, the name was changed permanently (Cavendish, 1999).

period 1944–1947 saw a brief return to a more democratic government, but two coups – in 1947 and 1951 – proved to be the start of a 22-year period of military government in the Kingdom.

1950s–1970s: US-led Industrialisation and the Formation of State-owned Utilities

The period after the Second World War saw Thailand transformed from a predominantly agricultural economy into an economy focused on industrialisation and the service sector, driven by renewed discourses of state-led modernity. The expansion of power plants and electricity networks played an important role powering this transformation, which was led by a military government with strong support from the United States. Development of the power sector was one of the key priorities, and the three energy utilities in Thailand formed during this period can be seen as the outcome of the new nationalistic discourse of energy-modernity. Increased levels of wealth, and technologies such as radio and television created new possibilities for territorialisation by the state and royal family. But, at the same time, they also sowed the seeds for alternative discourses of modernity.

During the 1960s, the United States became increasingly involved in Thailand and the region, mainly because of the Cold War, which divided Southeast Asia. Thailand, which was seen as a strategic ally by the US, became the cornerstone of anti-communist activities in the region. Girling (1981) and Baker and Phongpaichit (2014), writing about the relations between the US and Thailand as patron-client relations, claim they were 'more intrusive than anything Siam had experienced in the colonial era' (p. 139). The US also encouraged the coup by Field Marshal Sarit in 1958. His governments would lead the country for 15 years. During this time, Sarit heralded a new discourse of state-led modernity based on industrialisation and the institutionalisation of capitalism, through the encouragement of private investment, increasing and upgrading public infrastructure and welcoming foreign investment and import-substitution industrialisation (Hewison, 2006). According to Stifel (1976), 'Sarit grasped the opportunity of using an existing backlog of plans for rapidly expanding the economic infrastructure and establishing his regime with physical monuments of modernity' (p. 1,191).

During this time, Thailand's state-led modernisation project was supported by the IMF and the World Bank, increasingly focused on developing countries after the economies of Europe were back on track (Willis, 2005, p. 39). The World Bank spent a full year in 1958 analysing the country's economy and its institutions and published a report titled *A Public Development Program for Thailand*, which was basically a blueprint for Thailand's economic development in the next five years (World Bank, 1959). Some of the main problems identified by the World Bank mission included a lack of clear vision and planning; insufficient capital for investment in infrastructure and other parts of the public sector; and lack of human resources for development.

The World Bank's recommendations firmly linked the state-led discourse of modernity with the expansion of the energy sector through large-scale centralised projects. One of the key areas for public investment in the World Bank report was the energy sector, accounting for 2.8 billion baht or 22 per cent of the total capital expenditures (Figure 3.3). The hydropower sector was identified as one of the key and underdeveloped sources of electricity for the future. The main emphasis in the report was on the Yanhee (later Bhumibol) dam, which was completed in 1964 and reached its total capacity of 779 MW in 2002 (EGAT, 2012). At the time of the mission, the dam was already under construction and the Bank expected it to 'greatly stimulate many lines of economic activity, and will also directly improve the living conditions of a large part of the urban and rural population' (p. 9). Other key areas of investment in the power sector were small, 'interim' diesel generators, expansion of the Mae Mo lignite power plant, and the expansion and upgrading of Bangkok's electricity distribution system. The finance for these projects would come in the form of loans from the World Bank (44 per cent), the Thai government (30 per cent), from the Export-Import

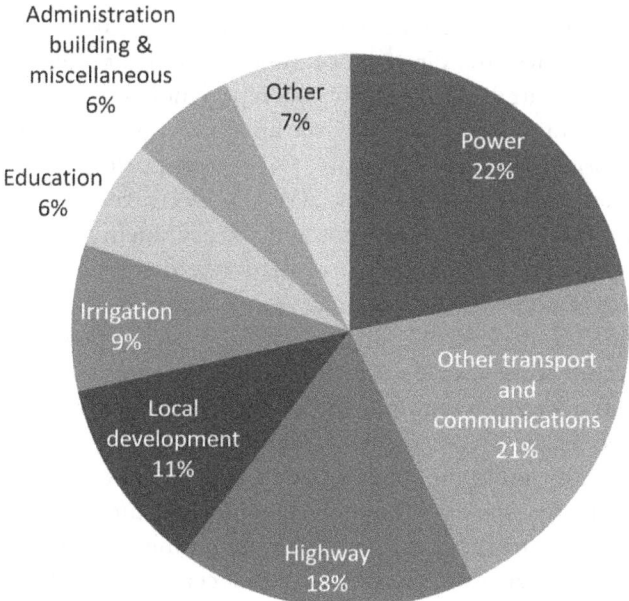

Figure 3.3   **Proposed share of capital expenditure on public development in Thailand from the 1958 World Bank mission report**

Source: Author. Based on a table in World Bank (1959, p. 25).

Bank and Development Loan Fund (later USAID) (23 per cent), and grants from the International Cooperation Administration (3 per cent) (World Bank, 1959). Because of these large investments, Thailand's expenditure on 'electricity and water supply' to GDP quadrupled from 0.4 per cent in 1960 to 1.8 per cent in 1970 (Hewison, 2006, p. 86).

Along with plans for infrastructure and other forms of public investments came ideas for institutional reform following the post-war trend of nationalisation of the utilities sector (Hausman et al., 2008; Williams and Dubash, 2004). In the 1950s, because the demand for power increased, the lack of strong institutions in the power sector became increasingly problematic. The generating capacity had increased five-fold from 40 MW at the start of the Second World War (1939) to 200 MW in 1958, half of it from an estimated 200 separate small cooperative, municipal or privately owned utilities (PEA, 2000, cited in Greacen, 2004). After the recommendations of the World Bank, a planning board, budget bureau, investment promotion machinery, and restructured central bank were set up (Baker and Phongpaichit, 2014). Important reforms in the power sector included the foundation of the Metropolitan Electricity Authority (MEA) in 1958, the uniting of the main power plants in the Bangkok area. This was followed by the establishment of the Provincial Electricity Authority (PEA) in 1960 and EGAT in 1969 (Sukkumnoed, 2007).

According to Greacen (2004), this period (1950s–1970s) was marked by a discursive struggle over the direction of rural electrification in Thailand. Both centralised and decentralised electricity generation models were championed by different state institutions. The newly-set up PEA supported grid expansion, while the National Energy Authority (NEA) supported the decentralised model. Greacen argues that the centralised model succeeded due to PEA's position under the Ministry of Interior and the alignment of centralised rural electrification with the state-led territorialisation and anti-communist strategy, which was supported by the United States. In contrast, the decentralised model was increasingly discredited and pushed into the background. The power of the NEA became eroded, evident in the frequent change of names, mission, and its position under different ministries of this organisation. Finally, in 2002, it was renamed as the Department of Alternative Energy Development and Efficiency (DEDE) under the Ministry of Energy (Greacen, 2004, p. 149).

Notwithstanding, the investments in infrastructure and the import-substituting policies had a strong effect on economic growth, amounting to an average annual increase of GDP of 6 per cent in the 1960s and 11 per cent in the 1970s (Hewison, 2006). Moreover, the urban economy itself was growing rapidly, a fact reflected in the change in Thailand's economic structure from predominantly agricultural to industry and services (Figure 3.4). In 1950, agriculture employed four-fifths of the total workforce and accounted for half of the GDP. By 1968, the share of GDP was down to one-third, on a par with

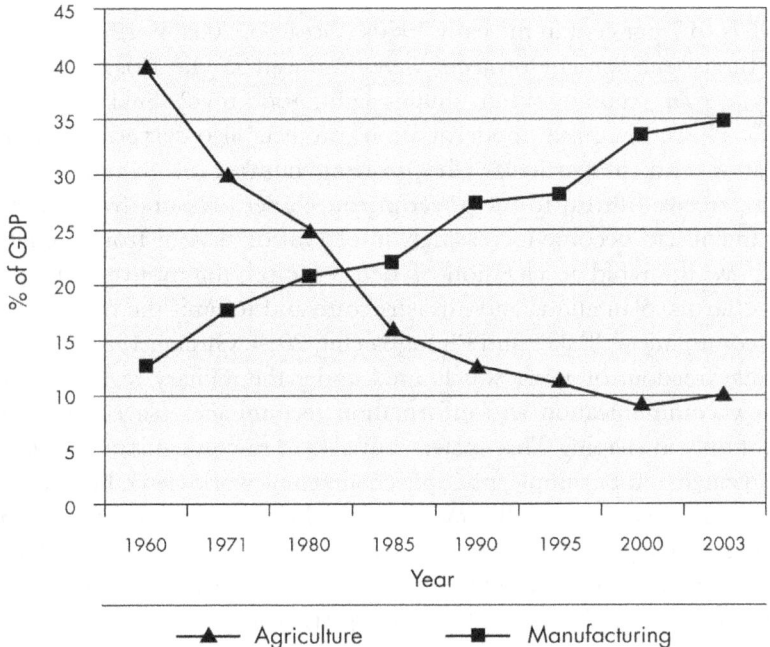

Figure 3.4 Change in the share of GDP for agriculture and manufacturing in Thailand from 1960 to 2003

Source: Hewison (2006, p. 87).

industry and services (Girling, 1981, p. 72). Most of the industrial activity was concentrated in and around Bangkok: it accounted for half of all the electricity demand in the country by 1958. The primacy of Bangkok was to increase in the subsequent decades. From the 1960s on, the tourist and expat industry developed along with the inflow of US experts and their families. Bangkok 'changed in shape, style, and tastes. New suburbs clustered around the schools, shops, cinemas, and clubs catering for westerners' (Baker and Phongpaichit, 2014, p. 148).

Rural areas also underwent important changes through the rapid expansion of infrastructure, and through attempts to include the previously remote countryside into the state-led discourse of modernity without a place for the communist 'other'. While the relative importance of agriculture diminished, in real terms the amount, size and productivity of farms continued to grow. Yields increased through the use of Green Revolution crops and technologies such as irrigation, pesticides, fertiliser, and the tractor ('iron buffalo'). While the Central Plains had been cleared and cultivated some time earlier, new farms were also being created in the North, Northeast and South of Thailand. US-sponsored highways, dams, airports and marine ports were constructed in these areas, and diseases such as malaria were eradicated (Baker and Phongpaichit, 2014). The rate of rural electrification rose from 2 per cent in the 1960s to 7 per cent in the early 1970s (Greacen, 2004, p. 127). Discursively, these rural development projects were captured in the slogan: '*nam lai, fai sawang, tang dee*' (running water, shining light, good road) (Sukkumnoed, 2007, p. 117). However, these 'modernisation' projects also served to control the population and, in particular, the growing number of 'communists', who were a perceived threat to the government. Under pressure from the US, the government had become increasingly intolerant of dissent from the left. This period saw the rapid acceleration of deforestation, the motive being to make way for farms, plantations and infrastructure and to limit the hiding places of these 'communists' (Baker and Phongpaichit, 2014; Girling, 1981).

While freedom of press was limited under the military regimes, electricity and new communication and information technologies played an important role not only in shaping Thai society, but also in re-confirming old hierarchies. TV and cinema, for example, not only created new worldviews, but were also an important means for King Bhumibol to raise his profile as the most visible and omnipresent King in Thai history (Baker and Phongpaichit, 2014). There was a strong link between the use of electricity and electric appliances and state-led development. As Williams and Dubash (2004) claim:

> In propaganda and popular consciousness alike, images of a society with universal and affordable electricity became important tropes of state-led development; the conjoining of the electrification enterprise to the majesty of the state can be seen in the expression of Thai peasants – fay luang, 'the king's electricity'. (p. 412)

Another way in which the king and the royal family were associated with electricity production and modernity was through the lending their names to some of Thailand's largest dams: Bhumibol Dam (1964, 749 MW), Ubol Ratana Dam (1966, 25 MW), Chulabhorn Dam (1972, 40 MW), Sirikit Dam (1974, 500 MW), Sirindhorn Dam (1971, 36 MW), and Vajiralongkorn Dam (1984, 300 MW).[3]

1970s–1990s: Thailand's Economic Miracle and Rapid Rural Electrification

From the 1970s to the 1990s, Thailand's plans for a centralised modernising state were briefly challenged by student-led coups, but they regained legitimacy under an essentially military government. Meanwhile, Thailand was on the road to become an export-oriented industrialised country. But, the increasing

---

3 The list shows the year of completion and the current installed capacity for each dam.

importance of business due to the inflow of foreign investment and linkages with the global economy were to prove a more lasting challenge to the power of the Thai elite and the bureaucracy. This also opened up space to challenge the centralised state-led discourse of modernity and its social and environmental impacts, through the proliferation of civil society organisations and NGOs. In effect, it marked the start of the environmental movement in Thailand. Furthermore, the benefits of economic growth started to spread to the provinces, by extension culminating in more political power in these areas, facilitated by improved transport, communication and information and driven by the spread of rural roads and increasing access to electricity.

The first part of this period saw a brief return to democracy in 1973 following a student-led protest. Both this protest and the coup of 1976 ended in brutal violence from the military and the eventual reinstalment of a military government. Meanwhile, the United States withdrew large numbers of its troops – and the aid projects that came with them – from Southeast Asia during this period after the end of the Second Indochina war (also known as the Vietnam War) in 1975. Their role was taken over by Japan in the late 1970s and 1980s, in terms of aid, loans and investment in electricity, water supplies, telecommunications, transport and agricultural credit. Finally, the oil crises of 1973 and 1979 had important repercussions for the energy sector and the country at large (Baker and Phongpaichit, 2014; Phongpaichit, 1980).

Substantial investment in the power sector as part of state-led energy-modernity, recommended by the World Bank in 1958, increased the share of loans of this sector to between a third and half of all foreign loans taken out. One of the main fuels for the power sector was imported oil. When the oil import bill tripled concomitant with the second oil crisis between 1978 and 1981, a debt crisis emerged in Thailand (Greacen and Greacen, 2004). One of the key consequences was the introduction of an export-oriented industrialisation model, following in the footsteps of the East Asian tiger economies (Hewison, 2006).[4] Moreover, the IMF demanded tariff adjustments and the privatisation of the public sector, notably the power sector. However, these demands were successfully contested by the country's labour unions and academics. Throughout the 1980s and 1990s, EGAT continued to invest heavily in generation and transmission capacity, financed by cheap bilateral loans (Greacen and Greacen, 2004). This eventually led to a decrease in the reliance on foreign fuel sources and a transition to hydropower and later gas-fired power stations. The Mae Mo coal-fired power stations also got a major upgrade. By the end of the 1980s, the share of oil for energy production had been reduced from 75 per cent to 20 per cent (Sukkumnoed, 2007) (see also Figure 3.5). By then, Thailand's power sector had become reliant on large-scale centralised power

---

4   Hong Kong, Singapore, South Korea and Taiwan.

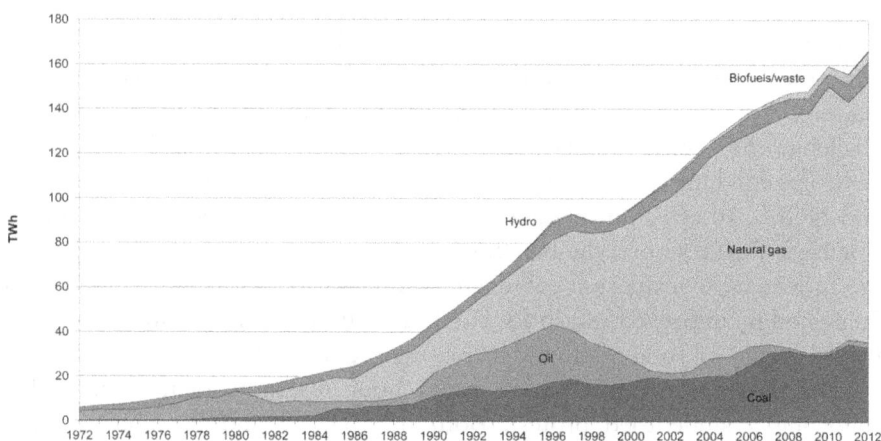

Figure 3.5   Electricity generation by fuel in Thailand from 1972 to 2012

Source: IEA Energy Statistics, www.iea.org/statistics © OECD/IEA [2014], IEA Publishing.

stations through EGAT's efforts to construct hydropower dams and gas-fired power stations. It was only during the oil crises that limited attempts were made to use renewable energy sources, that is, solar and wind energy.

The change to an export-oriented economy increased the pace of economic and social change in Thailand. GDP growth rates exceeded 16 per cent in the 1980s and were still in double digits in the 1990s. Contrary to the period of import-substitution, this strategy saw an increasing number of business people benefitting from these growth rates, resulting in the rise of a powerful new middle class (Hewison, 2006, p. 89). As Girling (1981) suggests, Bangkok became the absolute centre: 'Bangkok consumes more than 80 per cent of the nation's electricity, generates more than 80 per cent of its business taxes, holds more than 70 per cent of all commercial bank deposits, and absorbs slightly more than 60 per cent of the total annual investment in construction' (p. 88).

The middle class in Bangkok, along with a few urban centres, started adopting 'Western-style' consumption patterns, such as living in 'gated communities', shopping in large shopping malls and living more individualistically. An important driver of consumerism was television, which, by 1989, was being consumed by 80 per cent of all the people in Thailand (NSO, 2012). Around that time, there were also more than 400 radio stations in the country, and print media reached about two-thirds of all urban adults by the 1990s. Notwithstanding, during the 1980s the media content was still completely controlled by the military government and only slightly eased in the following decades (Baker and Phongpaichit, 2014).

While the balance of society and economy shifted from the rural to the urban areas, the former were not completely left behind (Baker and Phongpaichit, 2014, p. 199). Infrastructure development continued to bring 'modernity' to

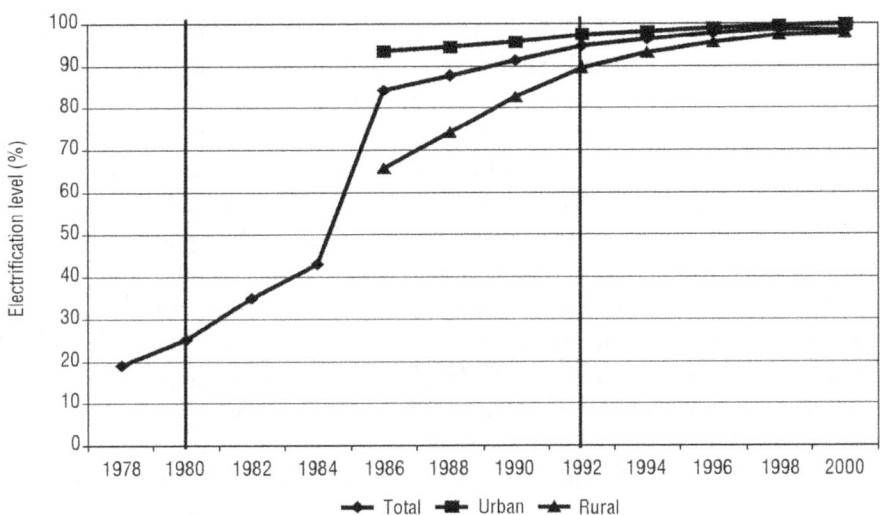

Figure 3.6 Electrification rate in Thailand for rural and urban populations between 1978 and 2000

Source: Reprinted from Shrestha et al. (2004) with permission from Elsevier.

rural areas, still driven by the desire to territorialise and control the countryside. The resultant increase in the electrification rate, both for the rural and urban areas, can be seen in Figure 3.6. By 1992, approximately 90 per cent of the rural and almost all urban households had access to electricity. The large investments necessary to fund this expansion made the Provincial Electricity Authority an increasingly powerful bureaucratic organisation. Moreover, the championing of their centralised model, which emerged during the Cold War, had by now dwarfed the few remaining decentralised approaches to rural electrification (Greacen, 2004).

The boundaries between the urban and the rural started to blur through the provision of infrastructure, diversification of rural livelihoods, and (seasonal) migration to urban areas. Baker and Phongpaichit (2014, p. 223) identify four entities that transformed the Thai village's relations with the outside world: paved roads, long-distance buses, television sets and Japanese motorcycles. While the roads, buses and motor cycles mainly reduced the physical distances in Thailand, television reduced the cultural and lifestyle gaps, and changed peoples' worldviews. By the 1980s, only a third of all rural households had television. Rural electrification and the extension of TV coverage increased this figure to 90 per cent by the mid-1990s (Baker and Phongpaichit, 2014, p. 224). After television, other household appliances followed such as fridges, fans and irons. As Greacen and Greacen (2004) contend: '[t]he material success of Thailand's electrification sent the message to Thai people that electricity is a public good to which Thai citizens, whether city dwellers or rural villagers, were entitled' (p. 519).

This period also marked the end of the hegemony of the state-led discourse of energy-modernity, challenged first by environmental movements and then increasingly by NGO and civil society opposition. Despite the country's strong economic growth, many people were still missing out on its benefits and their environment and livelihood opportunities decreased. The importance of agriculture declined, especially after prices dropped in the mid-1970s. As a result, some farmers opted to switch to high-risk activities, such as planting cash crops and shrimp farming, which sometimes created quick wealth, but sometimes resulted in larger debt burdens. In addition, many smallholders lost their land to large corporations, logging companies and dam builders. This led Girling (1981), after Jacobs (1971), to argue that Thailand was experiencing 'modernisation without development' (p. 84). This, along with the environmental impact of the country's modernisation project became the target of increasingly vocal academics, NGOs and civil society. The first NGO in Thailand was established in the late 1960s, but their number and influence increased from the 1980s onwards. The first major environmental movement targeted the Nam Choan hydro-electric dam in 1982, which would have flooded 223 km^2 of pristine forest in western Thailand. The project was delayed and finally cancelled in 1988 (Baker and Phongpaichit, 2014, p. 217). One of the responses to these movements by the military government and businesses was to become more active in supporting development projects in the Northeast, North and South of Thailand spearheaded by the King and his Royal Projects.

1990s–2010s: Mass Society, Partial Liberalisation and Alternative Energy-Modernities?

The period from the 1990s until the early 2010s was marked by not only further political and economic crises, but also continuing economic growth and demand for electricity. The partial liberalisation of the power sector opened up space for alternative energy discourses, such as regional power trade in Southeast Asia, and a renewed interest in decentralised forms of electricity generation. Furthermore, the establishment of the 'mass society' – the increasing economic participation and the compression of space through communication and mass media (Baker and Phongpaichit, 2014) – meant that the old form of centralised governance was further and more frequently challenged by mass protests and demonstrations. The ubiquity of television and the rise of the computer and the mobile phone which now reach virtually everywhere, continue to have an important homogenising, and therefore territorialising, effect in Thailand.

Politically, the last two decades ended a long period of military governments in Thailand, starting with the violent protest in 1992 which challenged the military dominance of Thai politics and the frequent military coups. This event reduced the role of the army and opened up political space for discourses of reform in the 1990s (Baker and Phongpaichit, 2014, pp. 253–7). The economy continued to boom in the early 1990s, driven by large amounts of foreign investment in manufacturing and increasing financial liberalisation. However, while this economic growth had increased the level of wealth for most people in an absolute sense since the 1950s, inequality had actually increased (Hewison, 2006, pp. 93–4).

In the energy sector, the discourse of neoliberal reform had unexpected consequences for the state-led energy-modernity discourse which had dominated Thailand since the late nineteenth century. The neoliberal discourse was different because it included ideas to create new competition for the state-owned utilities and potentially attract foreign investments. Thailand was one of the first countries in Southeast Asia to initiate neoliberal reforms in 1986 via the newly-established National Energy Policy Office (NEPO) and the National Energy Policy Council. For the power sector, the goals of the NEPO were to introduce Independent and Small Power Producers (IPPs and SPPs),[5] a market for electricity generation ('power pool'-model) and retail competition overseen by an independent energy regulator. All of these ideas strongly challenged the monopoly positions of EGAT, MEA and PEA, and their nationalist discourse of centralised electricity provision through large-scale power plants. Nevertheless, the first objective of reform was realised in the early 1990s, when private producers were allowed to generate electricity in Thailand for the first time since EGAT was established. Ironically, the first 'private' power producers were two of EGAT's own power plants, privatised under its subsidiary EGCO (Greacen and Greacen, 2004, p. 523).[6] The next step in the reform process was the introduction of the IPP bidding process in 1994, which led to the proposal to develop seven power plants with a total capacity of 6,680 MW. One of these proposed projects was the Bo Nok coal-fired power station. This project was successfully opposed by a local conservation group between 1995 and 2002 and transformed into a gas-fired power station at another location (discussed in detail as one of the case studies in Chapters 4 and 5). A second IPP round was held in 2008, with four more power plants with a total capacity of 4,400 MW to come online between 2011 and 2016. The SPP programme, which also started around 1994, has led to a total of 59 implemented projects producing 4,550 MW and selling over half of that to the grid by 2012. Figure 3.7 shows the transition which the IPP and SPP projects brought about

---

5 In Thailand, SPPs are power producers under 90 MW, whereas IPPs are all power producers larger than that.

6 EGCO = Electricity Generating Public Company Limited. After the initial public offering on the Stock Exchange of Thailand in 1994, EGAT's share was 40.7 per cent (Greacen and Greacen, 2004, p. 523). In 2014, its share was down to 25.4 per cent (EGCO, 2014).

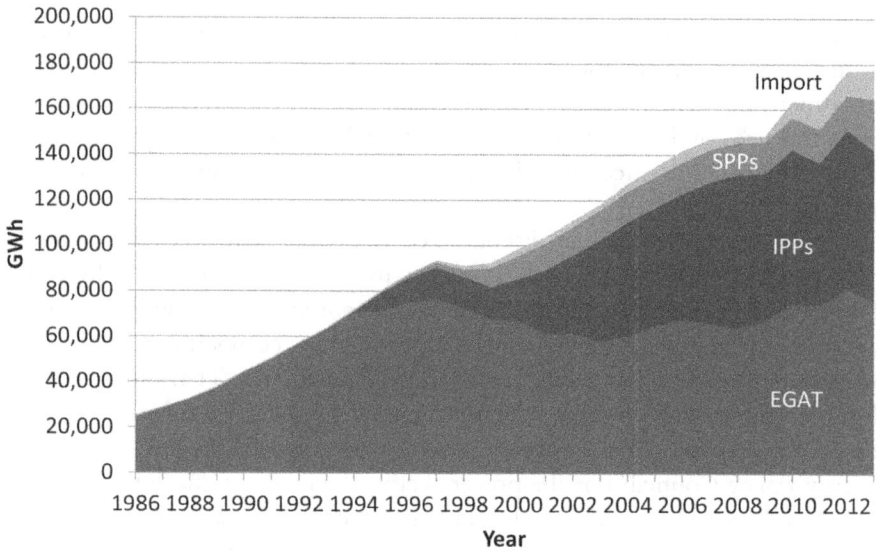

Figure 3.7   Power production in Thailand by EGAT, IPP, SPP and through import between 1986 and 2013

Source: Author, based on EPPO (2014a).

for the electricity sector in Thailand. Whereas before 1994, virtually all of Thailand's electricity was generated by EGAT, by the end of 2013 it was down to 43 per cent (EPPO, 2014a).

The 1997 Asian financial crisis unexpectedly and abruptly ended a period of four decades of volatile, but sustained growth. It also ended further liberalisation of the power sector in Thailand. Among the key causes of the crisis were: pressure on infrastructure, increasing wages, unrealistic property prices, speculation in the stock market, decreasing investment and overcapacity in a number of sectors (Baker and Phongpaichit, 2014; Hewison, 2006). The IMF stepped in again, this time with a US$17 billion package carrying a string of conditions, basically leading to more liberalisation and deregulation, increased privatisation, decentralisation, and fiscal austerity of the Thai financial sector. These conditions are widely believed to have deepened and worsened the crisis (Stiglitz, 2003), impacting not only businesses in Bangkok, but mainly the poor living in the urban and rural areas. Over seven million people fell below the poverty line of US$1.50 a day (Hewison, 2006). The IMF package also provided support for further liberalisation of the electricity sector and the setting up of a power pool, which would split EGAT, MEA and PEA into several companies and lead to full market competition. However, the ongoing crisis and its aftermath gave rise to popular opposition against these measures by a 'loose alliance of workers, intellectuals, NGOs, politicians, domestic businesses, and even the country's monarch' (Hewison, 2006, p. 98), which eventually overturned many of these reforms.

The aftermath of the economic crisis saw the emergence of a new kind of political landscape, epitomised by Thaksin Shinawatra, whose Thai Rak Thai (Thais love Thais) party won the 2001 and 2005 elections by an absolute majority. Thaksin proposed a populist policy programme designed to support rural villages, domestic capital and local businesses (Baker and Phongpaichit, 2014; Hewison, 2006). He halted further liberalisation of the energy sector and the three utilities (EGAT, MEA and PEA) were celebrated as 'national champions' under an 'Enhanced Single Buyer' model. At the same time, protests over the coal-fired IPP projects in Bo Nok (and nearby Hin Krut) reached their peak, exposing the high costs and risks associated with the model of privatisation in the Thai energy sector. The 'postponement' of these two projects[7] not only greatly discredited the pro-liberalisation NEPO office, but also led to the replacement of its director, and the renaming of the organisation as Energy Policy and Planning Office (EPPO) (Greacen and Greacen, 2004). Another key institute of the reform process, the Energy Regulatory Commission (ERC), was finally set up in 2007. However, its independence, capacity and political power have been questioned (Jarvis, 2008; Wisuttisak, 2010).

The rise of Thaksin's populist politics in Thailand, together with the successful protest movements against large dams and coal-fired power stations, further challenged what remained of the state-led discourse of energy-modernity. Two contributing factors were the gradual emergence of a 'mass society' in Thailand since the 1970s, and decline in influence of the bureaucracy and the military. The mass society in Thailand, according to Baker and Phongpaichit (2014), emerged as a result of the spread of infrastructure, education, the market, and the media: 'Information, images, and ideas arrived via satellite, TV transmission, film and internet. The economy became more exposed to global forces, and the society to global tastes and ideas' (pp. 233–4). Despite the above, the army staged another coup in 2006 following claims of corruption by Thaksin, who was subsequently exiled. What followed was a period of political turmoil and violence. The so called of 'yellow shirts', who were loosely aligned with the bureaucracy, palace and Democrat party, ranged against the 'red shirts', who were aligned with Thaksin's party and the lower socio-economic classes. The dissatisfaction with the Democrat party led to a landslide victory of Thaksin's party, led by his sister Yingluck, in the 2011 elections. Her government, in turn, was challenged from the start by the yellow shirts and increasing unrest and violence finally led to another military coup in May 2014.

---

7 Finally, the IPP-concession was used by the company to construct at 1468 MW gas-fired power station in Saraburi province.

An understanding of these democratic struggles and the increasing influence of civil society and environmental movement in Thailand is crucial to appreciate the increasing space for alternative energy-modernities in Thailand. As suggested earlier, an important starting point was the protest against the Nam Choan Hydropower Dam in 1982. The next big event was the first protest march in 1992, organised by the Northern Farmers' Network. This was followed by the Assembly of the Poor movement – formed through the protests against the Pak Mun dam – which was successful in getting concessions from the government in 1997 (Foran, 2007; Foran and Manorom, 2009; Missingham, 2003). Major successes of these movements were the postponement of the earlier mentioned coal-fired power stations in Prachuab Khiri Khan Province and the effective moratorium on all further hydropower projects imposed by EGAT after the ongoing protests against the Pak Mun dam.

The movements against the environmental impacts evolved into more specific movements and networks against the state-led energy policies and politics. Sukkumnoed (2007) describes the development and role of the Sustainable Energy Network for Thailand, which included organisations such as Palang Thai, the Healthy Public Policy Foundation, the Mekong Energy and Ecology Network (MEE Net) and the Alternative Energy Project for Sustainability. All of these organisations were active in the controversy surrounding the power plant in Bo Nok in one way or another and have since worked on various energy policy-related issues. For example, Palang Thai was instrumental in supporting the legislation of the VSPP policy and attempted to replicate this model in Laos and Tanzania. Unlike the civil society organisations and NGOs that focus on environmental issues only, the above organisations focus specifically on issues such as public participation in the Power Development Plan, decentralised generation policies, technical discussions about the reserve margin, demand growth predictions, and how to utilise intermittent (renewable) energy sources.

The emergence of these organisations and their protest not only resulted in the postponement of some large scale power projects, but also opened up space for alternative energy-modernity discourses. The first centred on the renewed space for decentralised electricity generation – much of which used renewable energy – as a result of neoliberal reforms and external discourses addressing climate change. This partly changed the model of centralised electricity generation and transmission adopted in the 1960s. Another important change was the introduction of feed-in tariffs – the amount dependant on the type of renewable energy used – for Very Small Power Producers (VSPP) in 2002 (<10 MW), which made it possible for the government to subsidise the sale of electricity to the grid for the next 10 years (Greacen and Bijoor, 2007). By March 2012, there were 264 such VSPP projects, generating 1,179 MW and selling

662 MW to the grid.[8] An additional 683 projects already had a power purchase agreement, but had not been built yet (EPPO, 2012). Three SPP projects received feed-in tariff, 11 received another form of government subsidy, and one project got both. While Thailand has seen various renewable energy plans, it is only since 2010 that the Alternative Energy Development Plan has been officially included in the Power Development Plan (EGAT, 2010), the main planning document for the power sector in Thailand.

Another increasingly important alternative energy-modernity discourse in Thailand is energy efficiency. EGAT has been involved in energy efficiency since 1993 through its Demand Side Management (DSM) programme, which has mainly been involved in energy labelling for different appliances and educational programmes. However, according to the director of this division, DSM is not considered a core activity for EGAT, since it is positioned under the Corporate Social Responsibility division. Moreover, the revenues of EGAT are still ultimately based on the amount of electricity sold (Foran, 2006). A more concerted effort towards energy efficiency is the 20-year energy efficiency plan proposed by the Ministry of Energy (2011), which aims to reduce the total energy intensity by 25 per cent in 2030 (compared to the 2005 baseline) and is included in the latest (3rd revision) of the Power Development Plan for 2012–2020 (Ministry of Energy, 2012).

A third alternative energy-modernity discourse in Thailand is the pursuit of nuclear energy. While nuclear power has been on the agenda in Thailand since 1966, it was only officially included in the Power Development Plan in 2007 (Bijoor, 2007). This PDP stipulated the building of five nuclear power plants, each with a capacity of 1000 MW. Besides the technical aspects, the activities of the Nuclear Energy Division of EGAT have so far focused on human resources development, feasibility studies, public consultation meetings, radio and TV documentaries, newspaper articles, press tours, exhibitions and seminars, awareness for EGAT staff, and educational material for citizens, including children's books and games (Boonpotipukdee, 2011). These efforts illustrate EGAT's awareness of the opposition against large-scale power developments and nuclear in particular. The accident in Fukushima (Japan) in 2011 has made implementation even more difficult and for this reason the proposed operation date of 2020 has been officially pushed back three years in the 3rd revision of PDP 2010 (Ministry of Energy, 2012), although the director of the Nuclear Energy Division at EGAT stated in an interview that it will be at least six years.

The final alternative to the national state-led energy-modernity discourse in Thailand is increased focus on imported energy from neighbouring countries. Since 2000, an average of approximately 22 per cent of the large demand for gas has come from Burma through two controversial gas pipelines (EPPO, 2013).

---

8 Not including the VSPP projects using fossil fuels.

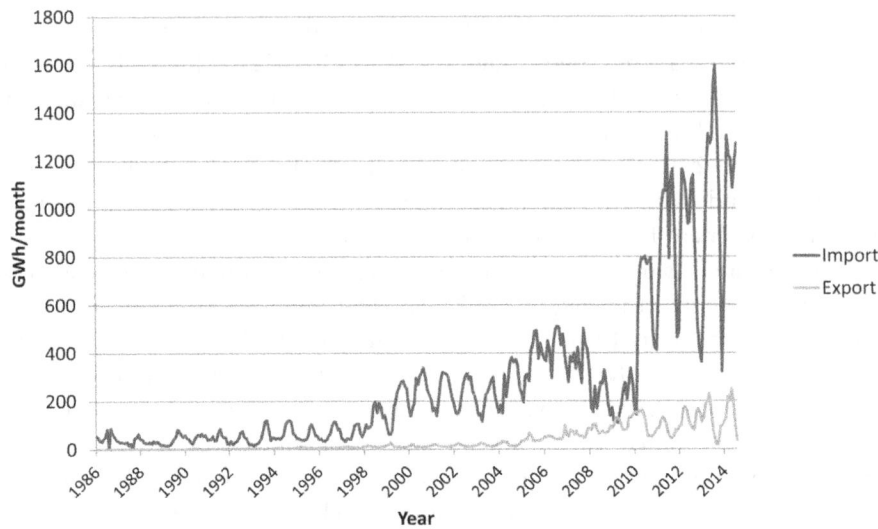

Figure 3.8   Import and export of electricity in Thailand until August 2014

Source: EPPO (2014b).

In addition, Thailand has been importing electricity from Laos since 1971. Subsequent to the opening of Lao economy in 1986, Thailand expressed renewed interest in the former's hydropower dams. It now holds a Memorandum of Understanding for the transmission of 7,000 MW by 2020 (Middleton et al., 2009). While the 1997 economic crisis put a temporary stop to these import-oriented projects, Figure 3.8 shows a steady increase in imported hydro-electric power (and its seasonal fluctuations), mainly from Laos. In an attempt to understand how Thailand's energy-modernity became increasingly intertwined with that of Laos through hydropower dam development and electricity import and export, it is this country that the chapter now turns to.

Modernity and Transitions in the Power Sector in Laos

1890s–1950s: Colonial Rule and the Seeds of Aid-Dependent Infrastructure Development

The end of the nineteenth century marked the first attempt at state-led territorialisation and institutionalisation of modernity in the area known today as Laos. Unlike Thailand, the current borders and foundation of the nation state of Laos were shaped by an external force, the French colonial empire. The overall impact of the French administration of Laos, however, was limited due to a general lack of interest and a relatively small number of officials. Notwithstanding, the French started to link the different remote provinces of Laos into their colonial administration employing a strong discourse of development and modernity. Laos' current dependency on foreign aid and foreign investment in infrastructure may be traced back to this colonial period (Phraxayavong, 2009), although only few roads and railways were built and the electric power sector was virtually non-existent.

While the earliest claims to Laos as a state date from the Lan Xang kingdom in the fourteenth century, the modern history of Laos starts with the treaty signed between France and Siam (Thailand) in 1893. Before that time, there were several different governed areas (*muang*) with claims over small parts of the country, and with little that united them. The treaty of 1893 fixed the boundaries between the colony of Indochina, of which Laos was a province, and independent Siam (Stuart-Fox, 1997). However, the French did not have much interest in developing Laos as a separate entity. There were at most a few hundred French officials throughout the country, the vast majority of them based in the urban and administrative centres. With limited manpower and limited possibilities to travel to the remote areas of the country, the French had to accept that many areas of the country would remain under their pre-existing governance systems, which in the remote rural areas where most of the ethnic minorities lived, meant no involvement of government at all (Scott, 2009). In effect, the French administration 'floated gently on top of the old native administration, giving an appearance of modernisation to what had already existed and with which it did not interfere' (Virginia Thompson, 1937, quoted in Evans, 2002, p. 47).

Despite these limitations, the French came with a strong developmental mission to govern 'what most considered the ignorant and benighted peoples of Laos. Villages were counted, censuses taken, headmen appointed' (Stuart-Fox, 1997, p. 31). Two central issues were the abolishment of slavery and taxation, but efforts were also undertaken to set up organised systems of justice and public works. The economy of Laos at that time was almost entirely based upon subsistence farming with some small exceptions in the urban centres. The benefits from tax and forced labour (corvée) were very low, covering only a small part of the costs of the French rule. As a side-effect, however, the lowland Lao peoples increasingly switched from a barter economy to a monetary economy. Furthermore, they also introduced the concept of national borders, cautiously promoted the Lao language, and, in that sense, contributed to the creation of the first 'modern' Lao citizens (Evans, 2002, p. 49).

Infrastructural development in the colonies, such as roads and (in particular) railways, was very important in the French colonial discourse and imagination of modernity (Starostina, 2010). While slow in Laos, the French did lay the foundations for some of its major roads. This slow pace may be attributed to the lack of profit from Laos as a colony, because the country 'raised just enough money to pay its officials and no more. There was nothing for development, road building, schools, hospitals, or any of the other fruits of the *mission civilatrice*' (Evans, 2002, p. 50). Only towards the end of the colonial period was some progress made on the development of infrastructure. Between 1930 and 1944, a few major roads were completed, some along the Mekong and others linking Vietnam and Laos. Many railways were planned for Laos and Indochina, but only one stretch was actually constructed in the south of Laos (Lee, 2003; Stuart-Fox, 1997). Evans (2002) summarises the infrastructural situation after the Second World War as follows:

> In the mid-1950s Laos had around 5600 kilometres of roads, of which around 800 were surfaced and therefore useable in the rainy season. In 1945 there were only nineteen registered vehicles in the country, a figure which had risen to around 100 by the early 1950s. The Mekong River and its tributaries constituted the main travel arteries, but only in some instances were boats driven by motor power. Air transport was minimal, and telecommunication was confined to the main centres. Telephone calls to provincial centres would not become possible until 1967 (Evans, 2002, p. 96).

While Laos had been a small and unimportant province of French Indochina for the first half of the twentieth century, the Second World War set the country up for a violent start to the second half. Early during the war, the French could not prevent Siam (Thailand) from 'reclaiming' some of the Lao territories they lost in the 1893 treaty. At the end of the war, Japan invaded Laos and drove out the French. When the former left again in August 1945, Laos was briefly independent until the French returned in 1946. By then, some sections of the Lao elite wanted to gain permanent independence, a position supported by the United States. However, it was not until 1953 that the French officially let go of Laos.

1950s–1970s: 'Secret War' and the Electrification through Development Aid

The period after independence from France until the Revolution of 1975 was marked by great political unrest, civil war and internal resettlement related to the wars in Indochina and the Cold War. However, this period also saw the construction of the first big hydropower dam (Nam Ngum 1) and the start of its export-oriented power sector, which marked the foundation of a discourse of energy-modernity based on centralised electricity generation in the country. Funding for this project was provided by the United States and its allies in the form of 'development aid', but it served a clear political agenda. The country's very small domestic electricity generating capacity, which was virtually restricted to the capital Vientiane, also relied heavily on development aid from the United States and Japan (Phraxayavong, 2009).

Almost immediately after gaining independence from France in 1953, Laos got caught up in both domestic politics and increasingly regional and Cold War politics, despite the pledge of the United Nations to maintain Laos' neutrality. One faction in Laos was supported by France, Thailand and increasingly by the United States; another faction, the Pathet Lao, and its political arm the Lao Peoples' Revolutionary Party (LPRP), had its roots in the Vietnamese 'Indochina communist party' and was supported by the Vietminh. Yet another group, which included long-time Prime Minister Suvanna Phuma,[9] tried to maintain Laos' neutrality by forming governments of national unity. But, the coalition governments established in 1957 and 1962 failed due to mutual animosity and the inability of the US and North Vietnam to reach a compromise. Finally, the LPRP was forced out of the government and the army and retreated to its base in the northern province of Huaphan. Meanwhile, the right-wing Royal Lao Government took over Vientiane and other areas in the Mekong valley. In 1964, all attempts at negotiation and neutrality failed, the country was divided into two parts and plunged into civil war, along with Vietnam. From May 1964, the 'secret' US air war started in Laos, concentrating on the strongholds of the Pathet Lao in the north-east and on the Ho Chi Minh trail in the southeast of the country (Stuart-Fox, 1997). According to Stuart-Fox (1997), the political struggles in the capital did not reach most of the citizens of the state, who continued to live as subsistence farmers:

> Unlike the Thai peasantry of the Chao Phraya valley, the lowland Lao had not become drawn into significant commercial production of rice, or indeed any other crop ... Lao peasants lived in a basically subsistence economy, in which families produced most of what they needed and bartered and traded for the few products they did not have, often with forest products. (pp. 95–6)

The legacy of the French electricity infrastructure in Laos was limited to just a few diesel generators in the capital city of Vientiane. Further development of the power sector would happen mainly through (inefficient) development aid. For example, the US aid mission in Laos sent three second-hand diesel generators to the capital in 1955, but did not provide any technical assistance. As a result, the generators could not be used for two years until the first French-trained Lao electricity engineer returned to the country (Phraxayavong, 2009).

---

9 Suvanna Phuma was prime minister of Laos from 1951 until 1975, with the exception of a few years in between. He was in charge of governments of national unity but also of the Royal Lao Government.

A few years later, the same engineer became the first director of Electricité du Lao (EdL), which was set up in 1959 and remained the first and only state-owned electricity utility in the country. The EdL website states that the company initially only serviced a few small diesel generators for a French base and a small part of Vientiane. The total capacity of the system at that time was just around 8 MW (EdL, 2011a), enough to power 8,000 households in the contemporary global north.

While in time Laos' dependence on the French government gradually diminished, the United States took over the French role of providing foreign assistance in dramatic fashion, in the hope of territorialising the areas under communist control. During the period from 1955 to 1963, 'US foreign assistance ... amounted to $192 per capita, the highest for any country in Southeast Asia ... By contrast, Thais received $31 per capita over the period from 1946 to 1963' (Stuart-Fox, 1997, p. 91). Most of this assistance was spent on military goals, such as the army, roads, transport and communications. According to Phraxayavong (2009) the US AID mission was so extensive that 'the US's influence often overshadowed and superseded even that of the Royal Lao Government. USAID even went so far as to create internal administrative divisions paralleling those of the Lao departments and ministries' (p. 119). However, very little aid went into support for agriculture, which meant that large numbers of the population were overlooked.[10]

Despite their large financial support, most of the development projects initiated by the US and its allies were not very successful regarding their objectives of territorialisation. Their projects bypassed large sections of the rural population, in particular the ethnic minorities in the upland areas under the control of the Pathet Lao. For example, US supported anti-communist propaganda only reached small numbers of the population, because of the high rate of illiteracy and the low rate of ownership of transistor radios (four sets per 1,000 people) (Stuart-Fox, 1997, p. 90). In contrast, the Pathet Lao were more successful in connecting and uniting those people living in remote parts of the country by working alongside the people. In addition, they 'set up 'mass' organisations in the villages of women and youth, and committees for seemingly everything' (Evans, 2002, p. 130). For some Lao people, it was the first time in history that they had become involved in issues of national interest and identity.

As a result of the large amounts of foreign aid and high levels of corruption, life in Vientiane and in a few other urban centres for the elite changed rapidly. Concomitant with the development of a small international community came the demand for international products, and more cars and motor bikes started

---

10 The amount of assistance received from the Soviet Union and Vietnam to the Pathet Lao during this period is unknown (Phraxayavong, 2009).

to be seen on the streets. There was a lot of demand for construction: theatres and other entertainment venues were established. Chinese entrepreneurs started investing in the small-scale production of food, metal and basic chemicals. As Stuart-Fox (1997) states, 'Lao society had become more sophisticated, more complex, more modern, but also more morally lax, more materialistic and hedonistic, more corrupt' (p. 155). Yet, despite – and in some instances because of – the war:

> communications throughout the country began to improve as telecommunications links were established between the main provincial centres and air transport expanded. This brought a more immediate sense of the physical expanse of their country to the Lao. (Evans, 2002, pp. 150–51)

The discourse of modernity associated with the development of large-scale hydropower in Laos emerged for the first time in the late 1950s and 1960s. Around this time, the allied-funded Mekong Committee (predecessor to the current Mekong River Commission) was set up to investigate the potential for cascades of dams on the mainstream Mekong and its tributaries – most of them in Laos – which can be interpreted as a way to territorialise the region and de-territorialise the increasing communist support (Bakker, 1999; Molle et al., 2009a). Despite the fact that many projects could not be built due to the high costs and risks involved, this discourse has nevertheless become central in the Lao economy, polity and international relations from that period onwards.

The first hydropower dams in Laos were the Selabam dam (5 MW) and the Nam Dong (1 MW), both of which were funded through French development aid in the 1960s. They became operational in 1970 (Phraxayavong, 2009, p. 98). The first large hydropower dam, the Nam Ngum dam, was built with technical advice from the Mekong Committee and the World Bank (Molle et al., 2009a). Construction started in 1966 and was financed by US$23.8 million from Australia, Canada, Denmark, France, India, Japan, the Netherlands, New Zealand, Switzerland, the United Kingdom and the United States.[11] Thailand supplied cement and electricity for the construction in exchange for future repayments in electricity. The dam therefore also marked the start of cross-border power trade in Southeast Asia.

The World Bank managed the funds in the first phase of construction of the Nam Ngum dam: the Asian Development Bank (ADB) took over in the second phase (World Bank, 1981b). Initially, only two 15 MW generators were commissioned in 1971, followed by two more 40 MW units in 1978. In 1985, the dam got a final upgrade, bringing the capacity up to 150 MW (World Bank, 1995).

---

11 Includes both phase one, finished in 1971, and phase two, finished in 1978. Most countries contributed to both, some only to one.

The dam also set a precedent in terms of social and environmental impacts; in total, some as around 800 families lost their houses and their agricultural land as a direct result of the construction of the dam. Yet, there was hardly any attention paid to – or recognition of – their losses because of the ongoing war (Hirsch, 1998, p. 63). Significantly, the export of electricity from the dam continued throughout the war and after the revolution, despite the political tensions between Laos and Thailand.

At the end of the 1960s and in the early 1970s, the US, now under domestic pressure, gradually started to withdraw its troops from Vietnam. However, the 'secret' bombing of Laos continued until 1973 when a new coalition government was formed. By that time, '[a]lmost all the 3500 villages under Pathet Lao control had been partly or wholly destroyed' (Stuart-Fox, 1997, p. 144). An estimated 400,000 or more people had lost their lives and a quarter of all people in Laos had become refugees in their own country (Stuart-Fox, 1997, p. 148). At this stage, the Pathet Lao, which was already in control of most of the country, took up key government positions. The economy had been hit hard by the decreasing amount of US aid as well as by the oil crisis of 1973, events that set the stage for a slow but effective coup by the Pathet Lao in 1975 (which became known as the Revolution). Around the same time, Saigon fell – ending the Vietnam War – and the Khmer Rouge took over in Cambodia. A few months after taking power, the Pathet Lao and LPRP forced the King to abdicate and proclaimed the Lao People's Democratic Republic (Lao PDR).

1970s–1990s: Socialist and Capitalist Experiments and the Start of Rural Electrification

From 1975 until the 1990s, Laos experimented with different economic discourses of modernity, socialist and later capitalist. Meanwhile, the country remained heavily dependent on outside support, both in the forms of economic assistance and knowledge and advice. Initially, the Soviet Union and Vietnam, which were the main partners and providers of foreign aid, tried to develop a form of modernity based on socialist principles (Evans, 1998). However, these ideas were soon watered down and progressively abandoned with the opening up of the economy from 1986 onwards. The export of hydropower electricity became one of the few sources of income and the prospect of more investment in this sector helped open the door to foreign capital and capitalist modernity in Laos.

In the early years after the 1975 revolution, several radical socialist experiments were introduced such as the nationalisation of industry and commerce and the cooperativisation of agriculture, albeit with generally negative results. Moreover, the people who had served under the Royal Lao Government were sent to re-education camps. As a result of these policies, approximately 10 per cent of the populace – mainly bureaucrats and Hmong but also 20,000 Chinese and 15,000 Vietnamese involved in trade and commerce – fled across the border to Thailand. Foreign oil companies were also driven out and fuel was rationed to minimise the import bill. All non-party organisations, newspapers and magazines were outlawed. The sum of these events damaged the administration, agriculture, and industry in Laos, and it slowed the reconstruction after the war. Soviet and other Eastern European aid could only partly replace the holes in the budget left by the United States (Evans, 2002). As a result:

> [t]he shutters came down on most businesses in the cities, cars disappeared from the roads as fuel prices shot up with the collapse of the economy, and bicycles reappeared on the streets in large numbers. Lipstick and make-up faded from women's faces, little jewellery was visible, and simple austere clothing became the unspoken rule. It was as if Vientiane had been suddenly propelled backwards into the 1950s. (Stuart-Fox, 1997, p. 177)

The state of the electricity sector after the Revolution was also pitiable, with the Nam Ngum dam heralded as the only exception. The World Bank (1981b) – whose involvement in Laos continued throughout the war and directly after the Revolution – estimated that the dam, generators, substations, and transmission lines were worth over 90 per cent of EdL's total assets in 1981 (p. 17). Despite their export earnings, the dams also provided electricity for the capital and surrounding areas:

> Electricity is readily available in Vientiane province and the present grid serves the city, the immediate vicinity, a few irrigation pumping units, and some villages around Vientiane and Nam Ngum. Roughly 20–25% of the households in the province have electricity and monthly consumption averages 130 kilowatt hours (kWh) per connected household (World Bank, 1981a, p. 10).

Outside of Vientiane, electrification was extremely limited in the early 1980s. Only four other provincial capitals had some access to electricity: two of which had to import it from Thailand (Savannakhet and Khammouane), and two others had small hydropower and diesel generators (Champassak and Luang Prabang). Imports (12–14 GWh/year) exceeded the amount of domestic production (8 GWh/year) in these provinces (World Bank, 1981a, p. 10). The EdL head office in Vientiane did not receive any information relevant to the costs and revenues from these power plants and imports, indicating the poor state of administration (World Bank, 1981b, p. 9). While foreign newspapers, magazines and videos were banned, people with access to electricity near the Mekong river could watch Thai television to keep in touch with the world outside of Laos (Stuart-Fox, 1997, p. 189).

The success of the Nam Ngum dam kept the discourse of energy-modernity based on hydropower in Laos alive. Indeed, one of the objectives of this dam, according to the World Bank (1981b), was to:

[a]ssist the Laos' Government and EDL in the study of hydroelectric schemes within Laos to establish which projects should be considered next for development with cooperation of Thailand which is fast running out of low cost renewable energy options. Laos, on the other hand, has many potential hydroelectric sites which could be developed for the mutual benefit of Laos and Thailand. (World Bank, 1981b, p. 10)

In 1979, the Nam Ngum dam contributed approximately 5.6 per cent (788 GWh) of the Thai electricity demand: the revenue from the electricity contributed around 35 per cent of the Lao oil import bill (World Bank, 1981b, p. 21). The transmission line of the dam was used to import electricity into Laos in the dry season, something that happened more often as the electricity demand in Vientiane grew (World Bank, 1981a). This was particularly significant as the borders between Laos and Thailand were otherwise officially closed and the political relations were tense. Even during the armed border disputes of 1984 and 1987, the export of electricity continued uninterrupted, as had happened throughout the last few years of the war.

As early as 1979, the strict economically Marxist discourses were loosened, setting Laos on the path to become a single-party government controlling a capitalist free-market economy. According to Stuart-Fox (1997), drought resulting in decreased electricity production from the Nam Ngum dam increased the urgency and speed of economic reform despite internal divisions (p. 199). This direction was formalised at the Party Congress meeting in 1986 at which the 'New Economic Mechanism' was introduced, following the examples of Vietnam and China. Measures included: (1) a move to market prices and resource allocation (including the currency); (2) a shift from central planning to guidance planning; (3) the elimination of subsidies and introduction of monetary controls; (4) the decentralisation of control to industries and lower levels of government; and (5) the encouragement of private sector and foreign investment (Rigg, 2005, pp. 19–20).

1990s–2010s: Open Economy, Accelerated Rural Electrification and the Hydropower Boom

From the 1990s on, the political situation in Laos stabilised and the economy became more open, following the NEM. The country has emerged as a new frontier for investment in hydropower, mining and other natural resources (Barney, 2009). Hydropower electricity has assumed a central role not only in the economy and international relations, but also in contemporary discourses of 'development' and poverty alleviation. Rural electrification, which has long lagged behind, has become a key target for the government to alleviate its least-developed country status,[12] supported by more development aid from more different countries than ever before (Phraxayavong, 2009). Moreover, the socialist experiments conducted after the revolution have been replaced by neoliberal and increasingly regional discourses of modernity. As Pholsena (2006) argues: 'As the country progressively opens itself to the market economy and to regional and international tourism, anti-capitalist and anti-Western imperialist rhetoric is no longer appropriate for galvanising the population behind the leadership. The discourse of struggle is being replaced by a discourse of lack' (p. 218).

Since Laos opened up its borders in 1986, its culture, society and economy have become closely intertwined with – and dependent on – those of Thailand. According to Evans (2002), 'Thailand's economic influence over Laos [since the late 1980s] is partly the outcome of the country's rapid growth during the halcyon days of the Asian boom, while Laos was still "building socialism"' (p. 226). A major event in Thai-Lao relations was the opening of the first Friendship Bridge between Vientiane and Nong Khai in 1994. Both materially and symbolically, the bridge across the Mekong River marked a change in the position of Laos from buffer state or backwater to becoming the crossroads of mainland Southeast Asia (Jerndal and Rigg, 2000; Pholsena and Banomyong, 2006). The opening of the border and easing of control also increased the penetration of Thai culture into Laos beyond television and radio. Tourists, newspapers, soap operas and music from Thailand 'brought with them images of the good life in Bangkok, including lavish lifestyles and conspicuous consumption' (Stuart-Fox, 1997, p. 205). Moreover, the Lao people are increasingly experiencing the discourses of modernity in Thailand first-hand, through seasonal or permanent labour migration (Phetsiriseng, 2001).

The discourse of energy-modernity based on the export of hydropower electricity, which had started in the 1960s, gained new impetus through the new discourses of capitalism, neoliberalism and regionalisation, following on from the reforms of the late 1980s.[13] Discursively, this was captured by Laos' ambition to become the 'battery of Asia' (Economist, 1993). As in Thailand,

---

12   The official criteria from the United National Conference on Trade and Development (UNCTAD) to graduate from the least-developed country status are a low-income criterion, a human asset weakness criterion and an economic vulnerability criterion. While electrification is not included in these criteria, it is nonetheless included in the tables of reports and generally seen as an important prerequisite (UNCTAD, 2011).

13   Indeed, Rigg (2005) argues that the measures taken under the New Economic Mechanism (NEM) have a lot in common with the standard neoliberal Washington consensus prescriptions for developing countries.

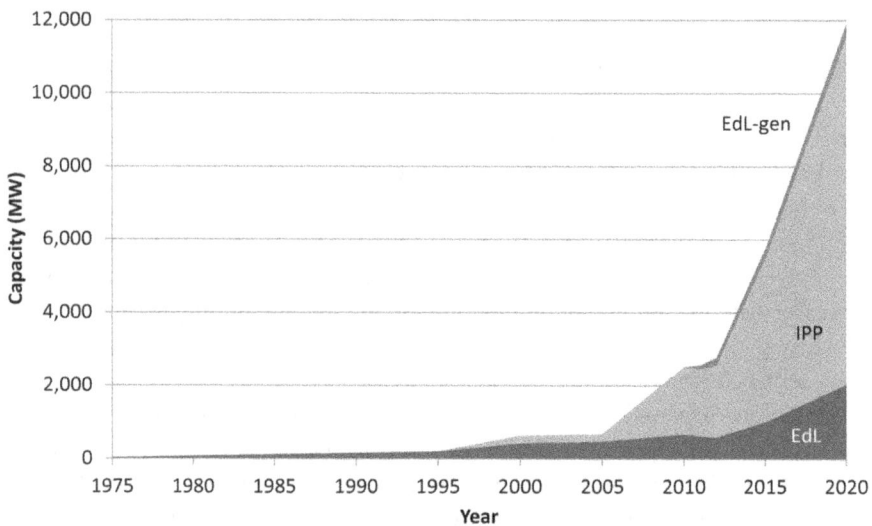

Figure 3.9   Electricity generation by EdL, IPPs and EdL-Generation from 1975 to 2020

Source: Author, based on data from EdL (2012a). Based on actual data for 1975, 1995, 2000, 2005, 2010 and 2011, and projections for 2012, 2015 and 2020.

IPPs territorialised large parts of this new space, by bringing in finance, expertise and human resources through 'Build-Own-Operate-Transfer' schemes (Wyatt, 2004). The first two IPPs to be completed were the 210 MW Theun-Hinboun (1998) and 152 MW Huay Ho (1999) dams, which, like virtually all other IPP projects, are foreign-owned and financed, with the Lao government typically holding 10–25 per cent of the shares (GoL, 2012). The influence of these and other IPP projects on electricity production in Laos may be seen in Figure 3.9. In 2005, EdL owned 69 per cent of the generating capacity, but despite a slight absolute increase in generation capacity, it dropped to 27 per cent of the total by 2010. This is projected to reduce even further to 20 per cent by 2020 (which includes 3 per cent by EdL-Generation) with 80 per cent of the capacity owned by IPPs.

The pace of hydropower development slowed around the time of the Asian financial crisis of 1997, but regained momentum in the 2000s. As of July 2014, there were 17 projects operational (GoL, 2014) and more than 70 under construction or consideration in Laos (GoL, 2012). All but one of these projects were hydropower projects (the 1,878 MW Hongsa lignite plant being the exception), with nine planned to be constructed in the Mekong mainstream in Laos. Figure 3.10 shows a graphical representation of the number of power projects, up until 2020, excluding those in the feasibility stage. Many of the developers, construction and consultancy companies are Thailand-based, and Thailand also provides one of the largest markets for electricity in Laos. China,

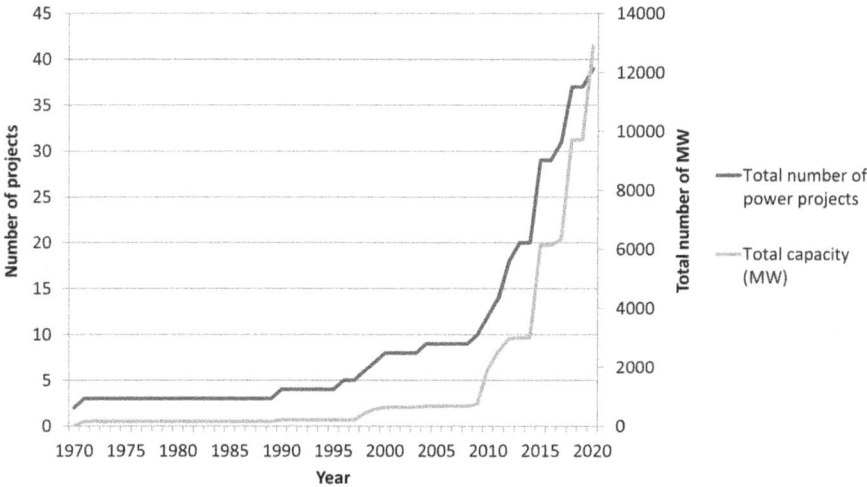

Figure 3.10 Total number and MW of actual, under construction and planned power projects in Laos per August 2012

Source: Author. Adapted from Government of Laos (GoL, 2012). The figure does not include the dams currently in feasibility stage.

Vietnam, Russia, France and Malaysia constituted other major investors and companies. In 2011, the GoL had signed Memorandums of Understanding (MoU) with Thailand (7,000 MW), Vietnam (5,000 MW) and Cambodia (1,500 MW), to be supplied by 2020 (Phomsoupha, 2009).

The increasing mutual dependence in terms of electricity adds another dimension to the social and cultural proximity of Thailand to Laos. Figure 3.11 shows the development of imports and exports in Laos. The key source of exports from EdL since 1971 has been the Nam Ngum 1 dam, but this has been overshadowed by the many IPPs that started exporting electricity in the late 1990s. At the same time, Laos has been importing electricity from Thailand to supplement its fragmented grid and compensate for the lack of electricity during the dry season. The import statistics show that import increased considerably since the late 1990s – from China and Vietnam also – due to increasing domestic demand in Laos.[14]

The state-led discourse of energy-modernity in Laos and the boom in hydropower, mining and other projects driven by foreign direct investment (FDI) has led to rapid development of material wealth, mainly concentrated in the urban centres. The number of cars, for example, has increased more than three-fold from 35,600 in 1990 to 121,800 in 2007 (ADB, 2012b), while the

---

14 The drop in 2011 was caused by reduced import from EGAT, probably due to increased domestic capacity.

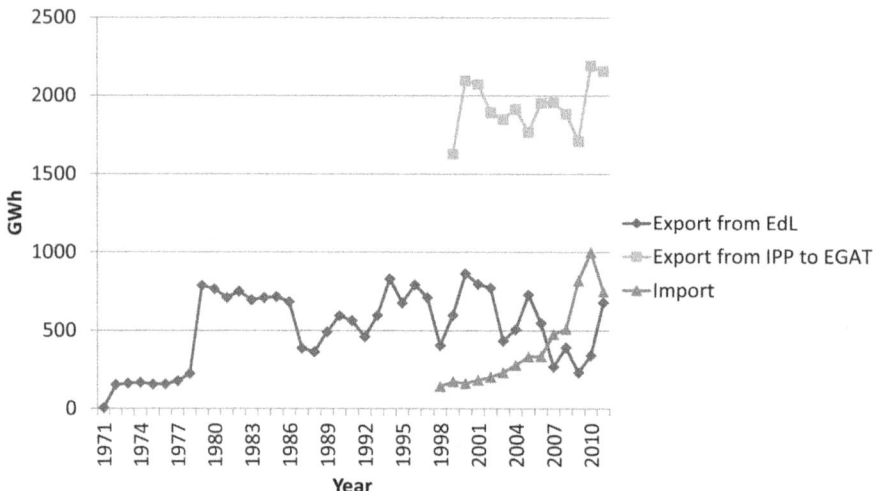

Figure 3.11 Electricity import and export by EdL and IPPs in Laos from 1971 to 2011

Source: Author, based on EdL (2012b). Import data only available from 1998 and IPP data from 1999.

number of mobile phone subscriptions increased almost 10-fold during the same period (ADB, 2012a). Vientiane, though still small, now has daily traffic jams, many international restaurants, a busy tourist industry and an international stadium (built for the 2009 Southeast Asian games). The construction of new roads and residential areas is rapidly expanding into its surrounding paddy fields. Therefore, as Rigg (2005) argues, '[t]he culture of modernity [is] propelled not only by government policies but also by traders and television and radio is creating a mental context where the products of modernisation become valued and sought after' (p. 31).

The discourse of energy-modernity based on hydropower has increased inequality in the country, both between rural and urban and rich and poor. The hydropower dams, built in the name of modernity and poverty reduction, have led to the resettlement of people from their land, increased illegal logging activities, and have greatly limited traditional livelihood activities such as fishing. Moreover, the costs of hydropower projects are often underestimated, while the profits are overestimated. This is both a general critique (e.g., WCD, 2000), as well as one that has been levelled at specific projects in Laos, such as the Theun-Hinboun dam (Barney, 2007; Shoemaker, 1998) and the Nam Theun 2 dam (Lawrence, 2009; Ryder, 2004; Singh, 2009). Finally, there are many questions surrounding how much of the profit goes to 'development' and how much ends up in the hands of foreign shareholders or a few people within the government (Bakker, 1999; Hirsch, 1998; Simpson, 2007; Wyatt, 2004).

From the 1990s on, electricity – in the form of rural electrification – became an important tool for state-led territorialisation under the discourse of development and poverty alleviation. Rural electrification in Laos had been neglected for many decades. The situation in 1992 was as follows:

> Provincial centres relied primarily on diesel generators, which are run for three to four hours nightly and serve only a fraction of the surrounding population. Most district centres do not have electricity other than small private generators that light the houses of a few dozen subscribers for several hours each evening. Automobile batteries and voltages inverters are used as a means of supplementing the limited hours of power. These devices enable Laotians to watch television and listen to stereo cassette players, even in remote locations. (Savada, 1994)

Laos' rural electrification ambitions are driven by the widely cited goal of providing electricity to 90 per cent of the households by 2020, which emerged in 2000 as part of a discourse to lift Laos out of the group of least-developed countries (LDC) by 2020 (UNDP, 2012). Like large-scale hydropower, grid expansion has de facto become the only technology to achieve domestic energy-modernity.[15] Since 1995, the number of electrified households has increased from 15 per cent in 1995 – basically only covering the main urban centres – to 78.5 per cent in 2011 (see Figure 3.12), financed by large (soft) loans and projects from the World Bank and the ADB, as well as bilateral loans from China, India and other countries (Bambawale et al., 2010). Reliability has improved, and electricity prices are still low compared to international and regional prices. However, since most of the district centres and areas along the main roads have now been electrified, further grid extension will be increasingly expensive as what one could term the 'low hanging fruits' have already been picked.

Laos' preference for grid expansion may be explained as part of a state-led territorialisation programme. Since 1975, the government has promoted a policy of so-called 'growth poles' or 'development centres' whereby rural communities are encouraged, and sometimes forced, to leave their remote upland areas and resettle in new villages further along the road (Baird and Shoemaker, 2007). The explanation provided by the government is that they are unable to provide basic needs and services to all small remote villages individually (for example, electricity, roads, water, schools and health centres). This relocation policy, which part of a wider programme of territorialisation and control, includes swidden agriculture, forests, water resources (Vandergeest, 2003) and electricity provision through the national grid.

---

15 In 2012, Laos did not have a centralised grid yet. Instead, there were three different grids: In the Central/North, Middle and South.

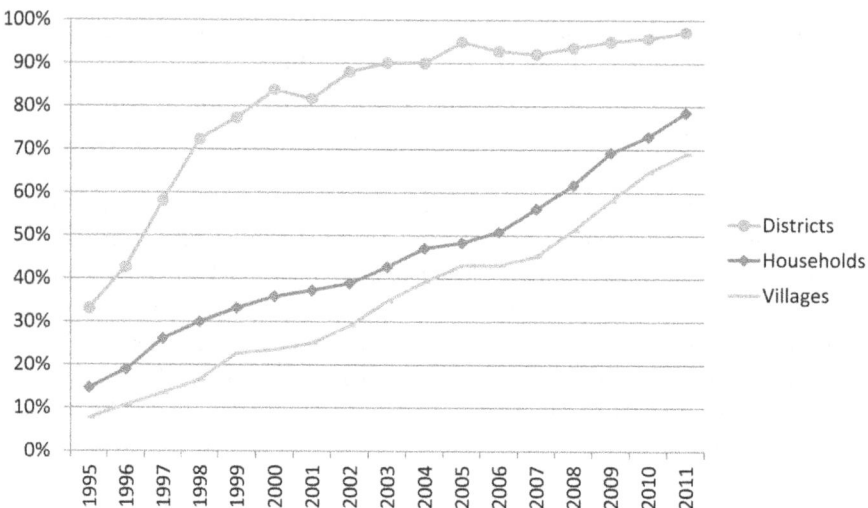

Figure 3.12 Percentage of districts, villages and households with access to electricity in Laos

Source: Author, adapted from data in EdL (2012b).

Off-grid alternatives to grid expansion – such as micro- and pico-hydropower and solar home systems (SHS) – have had limited success in Laos (Martin and Susanto, 2014). In addition, the objective of territorialisation, inter-governmental rivalry has also impeded better results, similar to the situation in Thailand (Greacen, 2004). Grid expansion falls under one of the largest state-owned companies in Laos, EdL, whereas off-grid is the responsibility of the Ministry of Energy and Mines (MEM). EdL, which has easier access to finance, prefers to implement large-scale electricity generation over off-grid or decentralised generation. MEM, on the other hand, which is dependent on short-term and project funding, often does not have enough capacity to implement these projects (cf. World Bank and AusAID, 2011). With the exception of SHS, which are supported by the World Bank under their Rural Electrification Projects, most off-grid systems are market-driven and not supported by donors or by government. In 2008, there were an estimated 60,000 pico-hydropower systems in Laos, providing electricity to approximately 90,000 households (Smits and Bush, 2010).[16] In comparison, only approximately 16,000 SHS have been implemented in Laos around that time, most of them fully funded by grants from the Global Environmental Facility (Susanto and Smits, 2010). Käkönen and Kaisti (2012) contend that many of the SHS fail and that despite 'the sense

---

16 Due to the expansion of the electricity grid since 2008, this figure is probably much lower now. However, more accurate data or estimates of the numbers of pico-hydropower systems in Laos do not exist.

of modernity and the social status that the solar panel provided' (p. 176), they often do not reach the most remote communities or poorest families in the villages. They find that SHS may even exacerbate poverty, as people have to sell assets (for example livestock) to pay their monthly fees.

The lack of alternatives to large scale hydropower for electricity generation and domestic rural electrification can also be attributed to the lack of space for civil society. Political freedom in Laos is still very restricted, and, unlike Thailand, there are no domestic NGOs or civil society organisations. Thus, as Evans (2002) argues, the international NGOs 'have become a kind of proxy Lao civil society' (p. 216). In Thailand, by contrast, decades of increasingly vocal and empowered civil society have led to more space for critique and the necessity to explain alternatives to state-led modernity and centralised power projects. Rigg (2005) similarly observes that '[c]ompared with neighbouring Thailand where there has been a long and sustained critical take on the fast-track industrialisation strategy pursued by successive governments, the picture from Laos is, on the whole, one-dimensional, lacking in both alternative narratives and nuance' (p. 24). In the closing paragraph of the revised edition of Evans's (2012) *A Short History of Laos*, the author comments:

> The hegemonic global mantra is 'development', an often vague term promising a better future, and almost anything can be justified just by invoking it. It is a kind of modern magic, and it trumps any other card in the deck, including preservation of 'a beautiful, ancient Lao culture', in the precious phrasing of the Lao Ministry of Information and Culture. Lao hope they can have their cake and eat it too; that they can have rapid all-round development that leaves their culture intact.

Explaining Regional and National Energy Transitions in Southeast Asia

The first part of this chapter analysed how bilateral arrangements pertaining to the power trade are becoming increasingly framed by discourses of regional cooperation in Southeast Asia. Within ASEAN and the GMS, cooperation is heavily influenced by the neoliberal polices of the World Bank and the ADB, with a specific vision of modernity informed by free trade and open economies. Despite the rhetoric associated with such policies, others have pointed out that these forms of cooperation may lead to uneven geographies of cost and benefit (Hirsch, 2007), by extension creating new inequalities (Glassman, 2010; Lawrence, 2009). In order to understand the development of these regional discourses, case studies of Thailand and Laos have been discussed in detail, showing not only their mutual dependence, but also two very different development trajectories within Southeast Asia.

The case study of energy, modernity and sustainability in Thailand demonstrated how the infrastructure and institutions of the power sector are shaped by domestic and international events such as the colonial period, the Indochina Wars, the Cold War and the global wave of neoliberalisation in the energy sector. Moreover, since the reign of King Chulalongkorn of Thailand, reforms in the country have been driven by a strong notion of modernity, expressed in administration, the development of infrastructure, and in power sector developments in particular since the 1960s. At the same time, this chapter shows how energy developments in Thailand are part of a specific form of energy-modernity characterised by increasingly critical stances against the notions of modernity propagated by the state and state-owned institutions. And while this criticism initially resulted in environmental and social movements against particular projects, it later develop into a more comprehensive critique of some of the core principles of energy policy and planning, such as lack of participation in decision-making, the neglect of externalities, and narrow focus on economic growth. Included among the – still modest – results of this critique of the prevailing discourse of energy-modernity are the proliferation of alternative energy sources (including nuclear power), increased public participation in energy policy and planning in Thailand, and increased focus on the import of energy from neighbouring Laos and Myanmar.

The case study of Laos has shown how this country – while sharing culture and language with Thailand – has developed in a very different way through its own specific history of colonisation, involvement in global politics during the Cold War, and finally ending up as a socialist government with a capitalist economy. Since 1986, however, Laos has reconverged with Thailand, partly due to their – discursively constructed – mutual interest in energy cooperation. Its smaller population, history of aid dependency, and limited space for civil society have not resulted in the same criticism of energy-modernity one sees in Thailand; rather, the entrenched culture of aid dependency has meant that this criticism is mainly voiced by international donors and NGOs. The result in terms of sustainable energy is a plethora of renewable energy plans and projects, but very shallow integration in governance structures of Laos.

To conclude, this chapter has offered a broad overview of transitions in the energy and power sectors in the context of political, socio-economic and cultural change in Southeast Asia. There are no single factors that can account for the many differences within and between countries in this region: they are the outcomes of changes at the local, national and global levels and their interactions. This showcases the complexity and non-linearity of energy transitions and the need to understand them as the outcome of processes on different scales.

In the next two chapters, I investigate four local energy trajectories to demonstrate and challenge some of the (implied) relations between energy transitions and local livelihood changes.

Chapter 4
Local Energy Transitions?
The Case for Energy Trajectories

Introduction and Key Argument

Where the previous chapter showed how energy transitions are embedded in processes of state-formation, territorialisation and modernity at both regional and national levels, the next two chapters look at how these transitions are embedded in livelihood change and modernity at the local level in Southeast Asia.

This chapter uses the concept of energy trajectories to analyse energy transitions on the local level. I define energy trajectories as 'analytical chronologies of the social, political, and material changes related to changing energy systems in a specific locality'. While there is some overlap with the concept of energy transitions, there are also some differences. As an approach, the concept of energy trajectories builds upon the understanding of the socio-technical nature of energy transitions, but does so for specific localities on a decadal time scale. In addition, energy trajectories give attention to the social changes related to energy, as well as to the social and power relations these technological innovations are embedded. As such, this concept also avoids attributing certain inherent qualities of scale or a given direction of interaction between scales (Marston et al., 2009; Neumann, 2009). Moreover, trajectories can include different energy end-use technology transitions, such as from biomass to gas for cooking, or from kerosene lights to electricity for lighting, which might happen for different households at different speeds and parallel to each other. Finally, the term 'trajectory' does not carry the teleological connotations of some conceptualisations of 'transitions', and stresses contingency, path dependency and non-linearity.

The key argument of this chapter is that state processes do not simply dictate what is happening at the local level; rather, there are different responses to the changes that take place, some of which are contradictory. As such, it deepens the argument that change in energy systems and livelihoods cannot be separated and that energy transitions are not a linear process dominated by state-led energy-modernity discourses. In addition to this argument, the chapter also aims to provide the reader with background information pertinent to the book's four case studies and their energy trajectories (see Figure 1.2 for their locations).

The rest of this chapter is structured as follows. Each of the four case studies and their energy trajectories will be discussed in sequence, starting with two cases from Laos and finishing with two from Thailand. As such, the chapter zooms in on the two very different neighbouring countries in Southeast Asia. In order to demonstrate the embeddedness of each of the energy trajectories, a substantial part of each case study in this chapter is devoted to understanding the history of each place and to the people's current livelihoods and how these have changed over the past few decades. I then move on to discuss details of the energy-related changes for all four energy trajectories. The chapter ends with a discussion of some of the key reflections on energy and modernity in each case.

Phakeo: Hybrid Village Grid in a Resettlement Area

The case of Phakeo (Laos) reveals some important dilemmas related to energy and development in the context of an upland community in Southeast Asia. Although the village was only established in 1998, it has attracted a substantial number of development projects since 2005, amongst them an NGO-funded hybrid-solar village grid. This case study also shows how the history of resettlement, the changes of connection to the road and markets, and the hybrid village-grid have influenced the local people's livelihoods. Furthermore, it is an example of the discourse of modernity promoted by the Lao government, which is often (implicitly) endorsed by the development community.

Geography, History and Development in Phakeo

Phakeo is a small village in Phoukhoun district, Luang Prabang province, Lao PDR. It is situated approximately 12 kilometres off the main road from Vientiane to the Northern provinces, approximately 15 kilometres from the district capital of Phoukhoun (Sam Nyek or simply Phoukhoun) (Figure 4.1). The district capital itself is a small sleepy town and the crossroads for people travelling from the capital Vientiane further north to the popular tourist destination of Luang Prabang and eventually to China, or to the west, to Xieng Khouang and eventually Vietnam. The area is at a high altitude and the landscape is mountainous. The climate is cool compared to the Mekong valley, with temperatures approaching zero during the winter.

Phakeo is situated in a resettlement area where, according to government officials, there were no settlements before the first village (Muang Chim) was established in 1993. Phakeo was the second village to be established in the 'village group' (*cum ban*), in 1998: there were 73 households in the village and 409 people in total (221 women and 188 men) by December 2010. Virtually

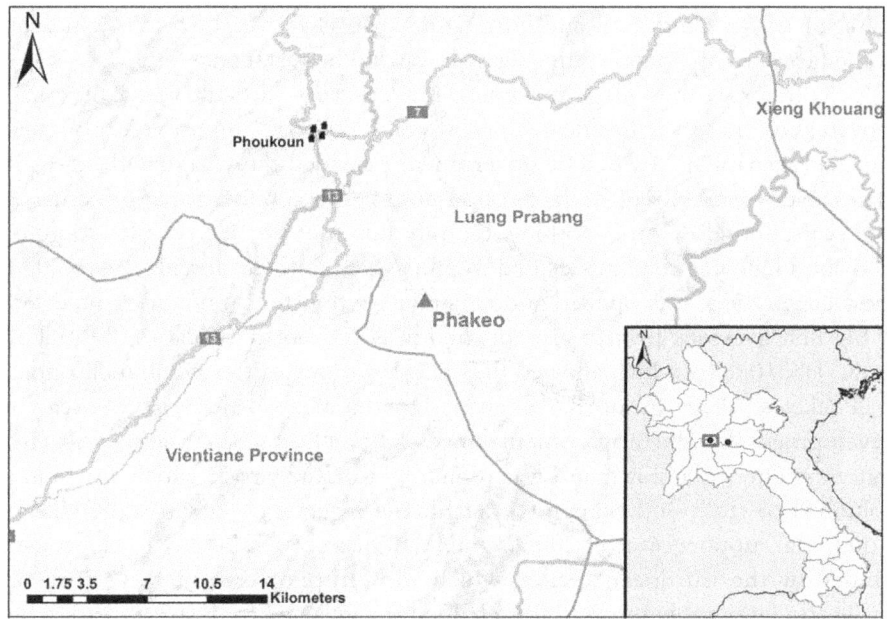

Figure 4.1 Map showing the location of Phakeo

Source: Author, adapted from Map data © 2015 Google.

all of the families that came to live in the village were of Khmu ethnic origin, one of the largest ethnic minority groups in Laos. They came from at least five different villages in four districts in Luang Prabang province: Daen (49 hh), Paklao (12 hh), Huay Hin Kham (5 hh), Pha Lak (3 hh) and Sam Nyek (3 hh). Most families moved in its entirety, although sometimes the grandparents would stay behind in the old village. The largest group, comprising those from Daen village, live at the north-western end of the village: the others are fairly spread out. Since 2005, no one has moved out of the village, because they would incur a fine from the government if they chose to do so.

According to the village group leader, the people were resettled to this particular area because it is devoid of dense or valuable forest. Moreover, it is suitable for keeping livestock, and there is plenty of land for paddy rice. The reasons the people themselves gave regarding resettlement were generally about wanting to find better and more convenient lives, built around government promises of improved food security and access to services such as education, health and electricity. Many people also mentioned wanting to improve the lives of their children through access to schools and opportunities to generate additional income to pay for school fees. The village headman acknowledged that resettlement took place in the context of the government strategy to eradicate poverty and slash-and-burn agriculture. However, he thinks that deep in the hearts of the people, they did not want to be resettled.

According to the village headman, life was very difficult when they first moved into the area. Besides the (seasonal) dirt road, there were no other facilities available in 1998. The government provided the village with tin roof plates, paddy land (about 2.5 ha per household), rice supplies for approximately two years, and also some clothing. Despite this support, the people struggled to make a living, and many decided to move back to their old villages. In 2000, the village school was opened and six gravity-fed water pumps were installed. A big first livestock project was initiated in 2003, worth a total of 80 million LAK (US$10,000), which allowed the villagers to buy cattle and animal fodder.

Phakeo's village group has been designated as an official target area for development by the government since 2005. During that same year, the water system broke down and was rebuilt by the Red Cross, which also sent a volunteer to study and change the health and hygiene practices in the village. Additional support came in 2008 in the form of a second big livestock project, funded by the European Union, which allowed people to buy livestock on credit (on favourable terms). This enabled the villagers to derive considerably more monetary income and led to strong growth of the village economy, according to the village leader. The livestock project coincided with the time when the hybrid village grid project brought electrification to the village and enabled people to buy electric appliances when the latter project was completed in December 2009. Another indication of Phakeo's increased economic prosperity was the number of vehicles in the village. In 2008, there were only one or two motor bikes in the village; by 2011 – according to the village head's estimation – there were between 30 and 40, including many tractors and one pick-up truck.

Phakeo has won a number of government awards, reinforcing its status as a target for development projects. In 2007, the village gained the 'cultural village' (*ban wattanatam*) and 'clean village' (*ban sam sa-at*)[1] awards. In 2008, it won the 'no crime' award, and an award for terminating swidden (slash-and-burn) cultivation. In 2009, the village was again recognised as a 'clean village', and in 2010 it again gained the 'cultural village' award. The signs to commemorate these achievements can be found at the centre of the village. According to the village head and other people in the village, these awards strengthened Phakeo's candidature for the hybrid village grid.

---

1 Literally, 'triple cleanliness'.

Changing Livelihoods

The livelihoods the people pursued prior to resettlement were different from their livelihoods today. Most of the people came from remote villages with few facilities such as roads, water, schools and electricity. The majority was used to growing upland rice, and to being almost completely self-sufficient. Some people mentioned that their former village is now more developed, through the construction of roads, schools, and the introduction of pico-hydropower electricity. Table 4.1 provides an overview of some key livelihood indicators for the 18 families interviewed in Phakeo.

Table 4.1   Key livelihood indicators for surveyed group in Phakeo (n=18)

Category	Indicator	Proportion of households (%)	Average	Minimum[1]	Maximum
Household size	Number of adults		2.89	2	7
	Number of children (<15 years old)		2.94	1	6
Agriculture	Paddy rice (ha)	94	0.57	0.14	1.50
	Upland rice (ha)	67	0.75	0.30	1.43
	Expense of buying rice (LAK/year)	44	1,950,000 (US$244)	540,000 (US$68)	5,500,000 (US$688)
	Income from gardens (LAK/year)	72	3,873,077 (US$484)	850,000 (US$106)	10,000,000 (US$1,250)
	Income from NTFPs (LAK/year)	39	5,750,000 (US$719)	2,500,000 (US$313)	10,000,000 (US$1,250)
Livestock	Number of cows	94	6.8	1	17
	Number of buffalos	56	4.8	1	14
	Number of pigs	61	4.2	1	12
	Number of chicken and ducks		16.1	2	40
	Income from livestock (LAK/year)	61	8,223,000 (US$1,028)	350,000 (US$44)	25,000,000 (US$3,125)

[1]   The minimum for all values higher than zero.
Source: Author.

While the livelihoods of the people of Phakeo are predominantly agricultural, they increasingly include off-farm activities. In the whole village, approximately 30 families were involved in wet-rice farming, covering a total area of approximately 30 ha. The paddy fields that were allocated after resettlement could not be used by all families, because, according to the village leader, only 30 out of the envisioned 100 ha could be irrigated. For this reason, some villagers decided to sell their land and revert to growing upland rice. This was found to be cultivated by most of the people in the village, including those with their own paddy fields.[2] The upland rice fields have been claimed and cleared by the people themselves. Only one of those interviewed had cleared new paddy fields: a few had bought or rented land from other villagers in Phakeo or in surrounding villages. The limited amount of land meant that not all families could grow enough rice to sustain them. Of the 18 households interviewed, only five reported growing enough rice to eat, 11 families had insufficient rice and had to sell livestock to buy more, and three families only had enough rice during years with good harvests.

Livestock was the villagers' main source of income, although this has varied a lot over the years; one of the difficulties being that people have often had to take their animals far away from the village for grazing, sometimes up to a day's walk away. As mentioned, various livestock projects have been implemented in the village, both by the government and by international donors such as the European Union. Most of these projects have made it possible for people to buy livestock on credit, with flexible conditions and lower interest rates compared to other ways of obtaining credit available to people in rural Laos. But, while raising livestock increases the opportunities for people to make money, it also constitutes risk of debt. Animal-related diseases are very common in Phakeo: many people reported having lost one or a few head of livestock over recent years.

All families were growing different kinds of vegetables, mainly for their own consumption and sometimes to sell to the market. The most common crops were corn, sugar cane, cucumbers, chillies, bananas and cassava. Many villagers were dependent upon the collection of Non-Timber Forest Products (NTFPs), such as orchids and other flowers, *kem*, bamboo and wildlife. Nine of the 18 families interviewed were involved in NTFP collection; seven realised a substantial part of their income from NTFPs, between 2.5 and 10 million LAK (US$313–1,250) per year. The actual value of NTFPs may be even higher given that many are for domestic consumption. The highest income from NTFPs, 10 million LAK (US$1,250), was reported by a family which buys said products

---

2   Note that upland rice does not necessary mean swidden or slash-and-burn cultivation, which was officially stopped in the village and recognised with an award by the government.

from Phakeo and the surrounding villages and sells them to middlemen. A 58-year-old man observed that nowadays children, who would not have engaged in making money in the past, are involved in the collection of NTFPs and are often better than adults at findings ways to earn an income. There are no fish in the rivers/streams near the village, signalling another major difference between the past and present livelihoods of many people.

While the livelihoods in Phakeo were mainly based on small-holder farming, there were signs that some production arrangements are changing. In 2010, Dao-Heuang company, a large and multi-sector company active in livestock production in the area around Phakeo, leased approximately 1,500 hectares of land from the government and expected to have 200 cows and 1,000 goats by the end of 2011. Company staff work with 10 families from Phakeo and a neighbouring village, who take turns to looking after the livestock and in return get 30 per cent of the profit. In addition, the Dao-Heuang company is also involved in growing organic vegetables such as cabbage and corn. The land concessions and new forms of contract farming arrangements, that have become increasingly common in Laos in recent years,[3] also reflect the spread of capitalist modernity (Baird, 2011). A 58-year-old man summarised this change succinctly when he says that before resettlement, life was comfortable (*sabai*) because there was rice, NTFPs and some livestock. People had enough to eat and did not have to spend much money. Nowadays, life is more convenient (*saduak*) because of the road, but it has also become more expensive.

Energy Trajectory

This extensive discussion on livelihoods allows for a better understanding of the energy trajectory of Phakeo, visualised in Figure 4.2. Before electrification, people mainly used kerosene lamps for illumination: some households would use up to 3.5 litres of kerosene per month. There were a few diesel generators in the village back then, which are still used for rice milling today. When the first family got their generator, the whole village would come to watch on their TV. This Family would charge 1,000 LAK (US$0.13) per adult and 500 LAK (US$0.06) per child to watch, as running such a generator would cost around 8,000 LAK/day (US$1). Later, when other houses also got diesel generators, these also shared their electricity with other households to reduce the costs.

---

3 Examples here are the so called 2+3 and 1+4 contract farming models. In the case of 2+3, the investors provide inputs, technical advice and access to markets, while the local people provide land and labour. In the case of 1+4, only the labour is provided by the local people (Fullbrook, 2007).

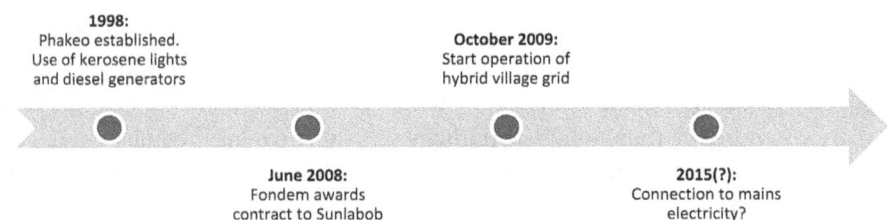

1998:
Phakeo established.
Use of kerosene lights
and diesel generators

October 2009:
Start operation of
hybrid village grid

June 2008:
Fondem awards
contract to Sunlabob

2015(?):
Connection to mains
electricity?

Figure 4.2   Selected events in Phakeo's energy trajectory

Source: Author.

In June 2008, Fondem, a French NGO, started the development of a hybrid village grid in cooperation with the Provincial Department of Energy and Mines (PDEM) of Luang Prabang. Fondem awarded the implementation to Sunlabob, a private Lao-based renewable energy company. The selection of Luang Prabang province was based on Fondem's previous experience in this province, and PDEM helped to find a target village for the project. Four years earlier, Phakeo had already been identified as a possible target for a project in a survey by Fondem.

Initially, the village authorities rejected the idea of a hybrid-solar village grid, because they wanted to be connected to the main electricity grid. However, when it became clear that this was not going to happen in the near future, they changed their minds. Other options, such as individual solar home systems, were dismissed by Fondem and PDEM because they were deemed to be 'more difficult' to realise. Yet, the costs for hardware and installation of equipment for the solar village grid (excluding costs by Fondem) were 105,000 euro (US\$136,000) in total or 1,460 euro (US\$1,890) per household. Given that solar home systems are widely used in Laos and only cost a few hundred US dollars each, one cannot but conclude that Fondem wanted to try out their model, regardless of the costs and experimental nature of the project.

Construction of the hybrid solar-diesel system started in August 2008 and its operation started one year later, in October 2009. The system consisted of 96 solar panels of 50 Watt each (total 4.8 kWpeak) which charged two sets of batteries during the day. The batteries were connected to two inverters, which were synchronised and fed into the village grid, connecting the individual houses. A 5.6 kW diesel generator served as back-up or battery charger in case of problems or long periods of limited sunlight (Figure 4.3). Sunlabob was contractually obliged to procure all of the equipment from Tenesol –

Figure 4.3   The Phakeo hybrid village grid system. Clockwise from the top-left corner: the solar arrays, the batteries, the diesel generator, and the control boxes

Source: Author.

a company partly owned by Electricité du France and one of the main partners of Fondem – which partly explains why the system was so expensive.[4]

The households were connected through a single-phase mini-grid. Each household was equipped with current limiter, which ensured that a household could not use more electricity – 8 W, 20 W or 80 W – than was paid for. The tariffs differed according to the amount of power available for each household (see Table 4.2). In 2011, 70 out of 73 households were connected to the hybrid village grid: the last three (new) houses were in the process of getting connected.

---

4   In comparison, the World Banks SHS in the Rural Electrification Programme phase 2 cost US$237 for a 20 W system and US$468 for a 50 W system (World Bank, 2009).

Table 4.2   Tariff options for users of the hybrid-solar village grid in Phakeo

	Number of lamps /socket	Monthly price (LAK/month)	Connection fees (LAK)
Service level 1 (S1)	1 lamp (8 Watt)	10,000 (US$1.25)	250,000 (US$31.25)
Service level 2 (S2)	1 lamp + 1 socket for radio or lamp (20 Watt)	20,000 (US$2.50)	250,000 (US$31.25)
Service level 3 (S3)	2 lamps + 1 socket for radio/TV (80 Watt)	30,000 (US$3.75)	250,000 (US$31.25)

Source: Author.

There have been a number of movements in each tariff level from the start of operation until it stabilised in April 2011. At the start, many people signed up for service level 2, but this share rapidly declined, because people initially thought that they could use the 20 Watt socket for a television and switched when they found out they could not. Service level 3, which allowed capacity for one television, was 'full' (as much as the design allowed for) almost from the beginning. The small slump that showed in the first few months was related to technical problems with the current limiters. After these limiters were replaced, the number of houses paying for service level 3 increased again to 18–19 households.

The tariffs were based on a willingness-to-pay survey undertaken before implementation and through negotiations with PDEM and the village authorities. They did not, however, reflect any actual maintenance and replacement costs. A simple calculation showed that for April 2011, a monthly sum of US$162.5 was collected, which would total US$1950 on a yearly basis if people stayed in the same tariff group. If all the money collected was saved for eventual replacement of the system, it would take nearly 70 years to collect the total capital and installation costs, not including wages for technicians and the village energy committee, diesel costs and occasional maintenance and repair costs. In other words, it was unclear what the money collected was actually supposed to cover. Moreover, there were persistent indications that the money collected by PDEM was not put into a dedicated account and was thus, in effect, 'lost'.

The system was supposed to provide light for four hours a day as contractually agreed upon between Fondem and the people in the village: one hour in the morning and three hours at night. For special occasions, such as weddings and festivals, the system could run longer using the back-up diesel generator. In practice, however, the number of hours the system could function depended upon the sunlight, the load connected, and the quality of the batteries. Since it was often cloudy in high-altitude Phakeo, the contractually agreed amount of electricity was often not reached. The daily operation and maintenance, and the collection of electricity fees and minor decision-making, were done by a village energy committee and village technicians. The main purpose of the committee was to oversee the system, the technicians, and to collect the monthly fees, whereas the technicians were involved in the daily operation of the system. Both the committee and the technicians had been trained by Sunlabob to carry out their tasks.

The hybrid village grid had not been running without problems. In August 2010, a lightning strike hit the system and damaged one of the inverters, which meant that the system had to run at half capacity or 2.5 hours per day. The issue took about half a year to resolve, because both Fondem and Sunlabob were reluctant to pay for the new inverter.[5] Meanwhile, the people still had to pay the full tariff, irrespective of the change of quality of electricity. In addition, a number of people were critical of the functioning and capacity of the system in general. A 58-year-old man said that he did not agree with the system from the beginning, because it was more expensive than grid electricity. A 32-year-old man said that he would be happy to change back to candle and kerosene light. Most people, however, seemed happy to have some form of electricity, but, immediately added that it was not enough. They would like to be able to use more than one or two lights and a few small appliances. Ideally, they would like to have grid electricity, like the villages on the main road. According to the Director of the Provincial Department of Energy and Mines of Luang Prabang, Phakeo and its surrounding villages would be connected to the central grid sometime around 2015. When this would happen, the hybrid solar-diesel village grid would be removed and possibly implemented in another village.

Reflections on Energy and Modernity

The history and energy trajectory of Phakeo provides some important insights into the co-evolution of infrastructure, energy and modernity on a local level. Initially, the village was drawn into modernity through the resettlement process and the various development projects introduced by the government and development actors. One 58-year-old man in Phakeo commented that, in the past, there was no development, but now there is development that comes from foreign aid. One of the preconditions for obtaining development assistance from the government – for roads, schools and electricity – was to move from the remote and more difficult places to areas closer to the main road. The GoL attempts to channel development projects, such as the hybrid village grid, from foreign donors into 'good villages', that is, into villages that comply

---

5 Note that they did not or could not use the funds collected from the monthly fees for reasons mentioned earlier.

with government's discourses of modernity. However, while many people who participated in the interviews said that they moved voluntarily, there were indications of elements of pressure as well, supporting Baird and Schoemaker's (2007) argument that resettlement is never entirely voluntary.

However, this case study suggest that resettlement has not been an entirely negative experience as Baird and Schoemaker (2007) and Baird et al. (2009) would have it. The interviews indicated that while the first few years were difficult, many people now seem to think of their new village as an improvement over their old habitat due to the road, the paddy fields, livestock, school, electricity and access to markets. One example was provided by a 33-year-old man, who said that he did not want to go back to the situation before resettlement because he wanted to look for work and benefit from increasing trade and development opportunities. Moreover, for the people in the village, electricity was also considered high priority and an important indicator of modernity in its own right. These observations are in line with those of Holly High (2008), who claims that '[r]esettlement taps into deeply held aspirations for poverty reduction and modernity amongst Lao rural residents' (p. 531). The limitations of the hybrid-solar village grid, however, left many people in Phakeo feeling disappointed. Such problems with the system in Phakeo were common among donor-funded rural electrification projects in remote areas of Laos; for example, the World Banks's SHS programme (Käkönen and Kaisti, 2012) and experiences with micro-hydropower, as shown in the next section.

Nam Ka: From Off to On-Grid and from Public to Private Micro-Hydropower

Nam Ka, another upland ethnic minority village in Laos, has a longer history of different electricity producing systems electricity. Its energy trajectory, which included the involvement of several development actors over almost two decades, provides another important example of how energy transitions and tensions of modernity are embedded in wider processes of changing livelihoods in Southeast Asia.

Geography, History and Development Projects in Nam Ka

Nam Ka, a village in Phaxay district, Xieng Khouang province, Lao PDR, is located approximately 35 kilometres from the provincial capital of Phonsavan, and around 10 kilometres from the district capital of Lat Kai. The map in Figure 4.4 shows that Nam Ka is situated in a valley which is an offshoot of the plateau on which Phonsavan is located. The plateau itself is approximately 1,200 metres above sea level and the landscape is hilly and dry, with only a

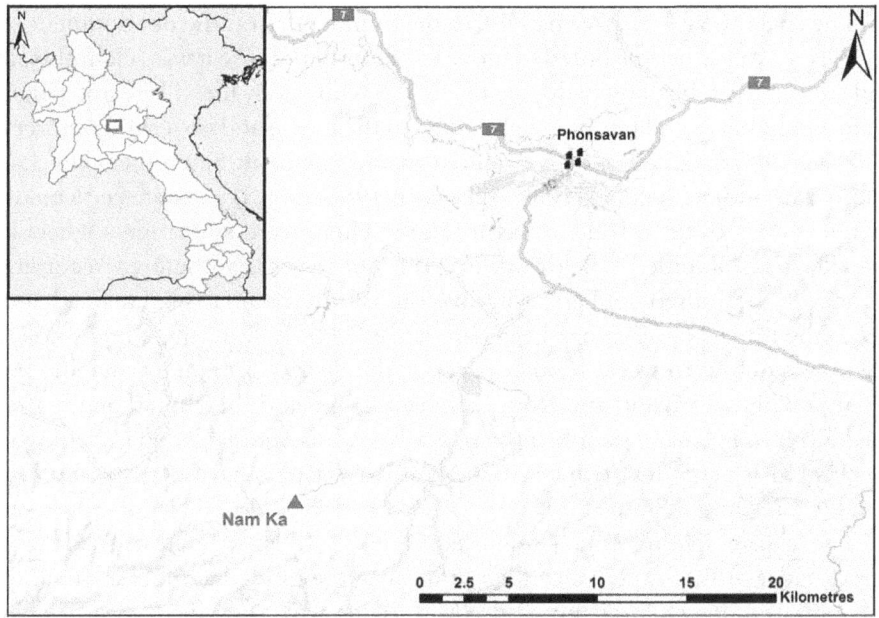

Figure 4.4   Map showing the location of Nam Ka

Source: Author, adapted from Map data © 2015 Google.

few patches of forest. Because of these characteristics, and due to its strategic location between Huaphan (the base of the Pathet Lao) and Vientiane (the seat of the Royal Lao Government), this was one of the strategic areas during the 'Secret War'. Heavy fighting took place on the plateau, of which in particular the use of cluster bombs is still a major problem. The village Nam Ka, which lies in a valley, is part of a more mountainous landscape on the same altitude as the Xieng Khouang plateau. Because of the altitude, the climate can be cold in winter, often nearly reaching freezing point. The summers are relatively cool, especially compared to summers in Vientiane, the capital of Laos.

While people had previously settled in the area, Nam Ka was officially established in 1982 in the aftermath of the civil war in Laos, when six ethnic Hmong families settled there. Around 1985 and 1986, another 64 Hmong families moved into the village from Nonghet district, near the border of Vietnam. At that time, there were two settlements or hamlets located approximately 30 minutes walking distance from each other. Over time, the village grew gradually, from natural growth as well as from further in-migration. Some families moved in and then moved out of Nam Ka in the first decade after its establishment, because they were afraid of the rebel groups who were roaming in the areas near the village. By 2011, all of the people in the village were from the Hmong ethnic group (totalling 112 households and 735 people).

The civil war and its aftermath had a major influence on the development of the village.[6] Many people opted to move to Nam Ka because it was relatively safe and suited to establishing paddy fields. In 1985, however, the village got caught up in a fight between Hmong rebels hiding in the jungle and government forces. One person was killed and five injured in this incident. Since then, the Lao army has stationed soldiers in the village. In 1987 the two separate settlements merged and moved to the current location. There were still some soldiers in the village at the time of fieldwork in 2011, but fewer than before. According to one of the retired soldiers, only five out of the original 14 remained, and he claimed that Nam Ka has been safe since 2009. As a result, some soldiers are now required to work in other parts of the district. Because many Hmong families fled to Thailand and from there to Europe and the United States after the 'Secret War', many people in Nam Ka have close family living abroad who come to visit every few years. A few Nam Ka villagers have even paid return visits to the United States.

The village school, which was established in 1984, offered only the first three grades at that time. Since then, it has been upgraded a number of times, and now teaches children up to grade 5. After that, they have to go to the lower secondary school in a nearby village (Tang Xiang Neua) if they want to continue their education. While the school employs both Hmong and Lao ethnic teachers, in practice there is still a language barrier. Many older people, particularly women, cannot understand the Lao language, and some children and adolescents struggle to speak the language, making it more difficult to engage with the world outside of their village.

Like Phakeo, Nam Ka has also had its share of development projects. In 1993, a logging company came to the village and built a road to enable it to remove the timber. The logging company tried to establish good relationships as well, for example, providing free light bulbs after a micro-hydropower plant was built by a Chinese company with funds from the Chinese state press agency in 1995. According to a 28-year-old man, neither the government nor other projects showed much interest in the village prior to the year 2000, with the exception of the logging company and the company that built the micro-hydropower system. After 2000, projects slowly started coming in: JICA (Japan International Cooperation Agency) built a school; the International Fund for Agricultural Development upgraded the road, built an irrigation weir and provided support for livestock; the World Food Programme provided rice; and another organisation set up a gravity-fed water supply. The Swiss NGO Helvetas started work on its Rural Income through Sustainable Energy (RISE) project in 2008, after it supported the rehabilitation of the micro-hydro system.

---

6   During the war, many of the Hmong were recruited by the US CIA to side with the Royal Lao Government. Others fought on the side of the Pathet Lao.

Helvetas has been involved in providing information, farming techniques, and supporting livelihood development activities such as handicrafts and some basic infrastructure. Initially, the focus of all activities associated with the project was on Nam Ka, but it gradually extended to other villages and districts in Xieng Khouang Province, including to advocacy activities at the national level.

Changing Livelihoods

The main livelihood activities in Nam Ka were upland and paddy rice farming, keeping livestock, collecting NTFPs, and growing vegetables such as chillies and papaya. Many of these activities were produced for own consumption. Similar to Phakeo, growing paddy rice and keeping livestock was relatively new to the people of Nam Ka. Originally, many people cultivated upland rice and supplemented their diets by wildlife and other NTFPs. Some key livelihood indicators of the 16 surveyed households in Nam Ka are shown in Table 4.3.

Table 4.3   Livelihood indicators for surveyed group in Nam Ka (n=16)

Category	Indicator	% of hh	Average	Minimum[1]	Maximum
Household size	Number of adults		3.69	2	9
	Number of children (<15 years old)		4.00	1	8
Agriculture	Paddy rice (ha)	81	0.89	0.2	2
	Upland rice (ha)	31	0.71	0.55	1.5
	Income NTFPs (LAK/year)	31	1,052,500 (US$132)	60,000 (US$8)	3,500,000 (US$438)
Livestock	Number of cows	100	6	2	14
	Number of buffaloes	100	2	1	15
	Number of pigs	94	2	1	7
	Number of chickens and ducks	100	7	3	25
	Income from livestock (LAK/year)	56	4,778,000 (US$597)	2,000,000 (US$250)	10,000,000 (US$1,250)

[1] The minimum for all values higher than zero.
Source: Author.

Table 4.1 shows that most of the people interviewed were involved in paddy rice production: around one-third grew upland rice and a small number grew both. Families who had moved in earlier could claim paddy fields freely and close to the village. For most households, however, the fields were three hours walking distance from their new habitat and they had to buy the land. A family who moved into the village in 1991, for example, bought 1.1 ha of paddy field for US$1,800. Families that moved in later, and did not have enough money to buy land, had to practice upland rice farming. Take, for example, the family of a 63-year-old man who lived with his wife (50 years old), son (26) and daughter-in-law (17). Although they first moved to Nam Ka in 1986, they had to move back and forth between their old village a couple of times before they permanently settled in Nam Ka in 2003. By that time, the family's only option was to cultivate upland rice on land which was approximately two hours walking distance from the village. The village seemed reasonably food secure, with only three families in the sample claiming to have insufficient rice almost every year, and one family for some years.

Livestock was the main source of income as well as an occasional source of protein. All of the people in the sample kept some cows, buffaloes, pigs (and poultry), although some households kept just one or two animals. In addition, there were many problems with diseases such as infections and foot-and-mouth disease. According to statistics collected by Helvetas (RISE, 2009), in 2008, almost as many buffaloes, cattle and chickens died of diseases as were sold or eaten. For pigs, the number of deaths was actually higher than those that survived for sale or consumption. The high incidence of disease explained why just over half of the people surveyed reported any income from livestock. Difficulty in finding land was also a problem in terms of livestock, as the land for grazing was becoming increasingly scarce. Some families had to walk a full day (eight hours) to the place where they kept their livestock. For this reason, groups of families took turns in sending people to look after the cattle.

The importance of NTFPs seemed to be in decline as far as the livelihoods of the people in Nam Ka were concerned. Table 4.3 shows that less than a third of the people of Nam Ka gained economic benefit from collecting NTFPs, such as bamboo, mushrooms and rattan. According to a 45-year-old widow, NTFPs were harder to collect than they were in the past. A 31-year-old female said that her family used to be able to earn a lot of income from wildlife, but not anymore. Another family said that they simply no longer had time to collect NTFPs, because they had a lot of land to take care of.

Similar to the case of Phakeo, the road and access to vehicles have increased trade and economic activities outside of the village. Increasing numbers of shops are selling food and consumer products brought in from the district or provincial capitals. Foods that used to be carried by hand now come by truck. Besides selling livestock, the people in Nam Ka also tried to sell corn and other vegetables to the market, but without much success due to the limitations set by the district authorities that stipulated that they could only sell to the district government and not to middlemen. Nonetheless there were more opportunities to make money, according to a 35-year-old man, as a result of better connections to the market. One of the deputy village leaders said that although it is easier to buy and sell things, life has also become more expensive.

Energy Trajectory

The timeline of Nam Ka's energy trajectory is shown in Figure 4.5. Before electrification, the villagers depended mainly upon kerosene lamps and burning pinewood for illumination in the evenings, although there were also one or two diesel generators in the village. In 1995, only a few years after the road was completed, the Xinhua News Agency Development Corporation[7] built a number of micro-hydropower plants in Laos. Nam Ka village was selected as a site by the government, along with four or five other villages in the province (JICA, 1999). Nearby, the Nam Ka 2 micro-hydropower project (81 kW) connected four of the neighbouring (ethnically) Lao villages. Unless otherwise stated, this book refers to the system that provided electricity to Nam Ka village. This was a 12.5 kW micro-hydropower plant that was connected to all of the households in Nam Ka, free of charge, to provide four hours of light every evening. The monthly electricity fee was set by the villagers themselves at 1,000 LAK (US$0.13) per household per month per light bulb, approximately twice as much as grid electricity. The Chinese company trained two people from the village to provide maintenance and the fees were used to cover the cost of maintaining the generator insofar as possible with limited knowledge and access to spare parts. After the generator broke down in 2003, some households bought diesel generators and pico-hydropower turbines. However, the majority of the people had to revert to using pinewood and kerosene lights.

In 2006, the system was selected for rehabilitation by a public-private coalition of Sunlabob, the same company that implemented the system in Phakeo; Helvetas, an NGO; and Entec, a Swiss-based engineering company.

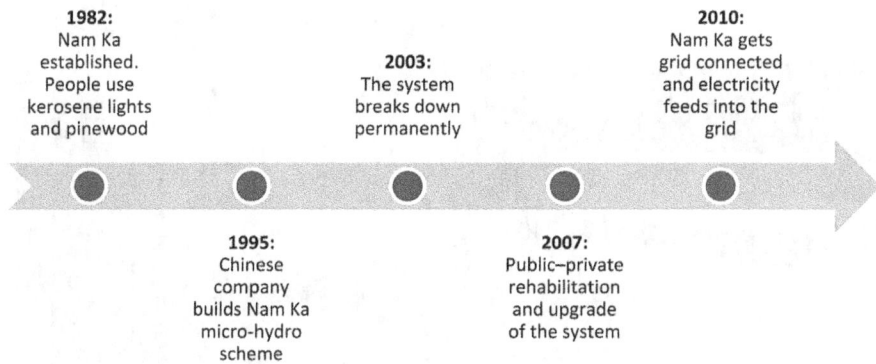

Figure 4.5   Selected events in Nam Ka's energy trajectory

Source: Author.

---

7   Part of the state press agency of China.

Most of the infrastructure, except for the small concrete dam, the forebay and the penstock, were replaced and a new generator and new grid infrastructure were set up (Figure 4.6). Moreover, the system was upgraded to a hybrid village grid, which included an array of solar panels (1.84 kW) and a back-up a diesel generator (15 kW). Sunlabob, which had developed this 'public-private partnership' concept – for which Nam Ka was the pilot or showcase project – coordinated the project and provided half of the funding for the 'moveable assets', the new generator and related infrastructure (approximately US$69,000). Entec provided the equipment, the software, the technical expertise and the other half of the funding for the generator, also around US$69,000. Helvetas provided approximately half of the funding for the 'public assets' (US$20,000), such as a new village grid. The remainder of the cost of grid infrastructure were met by Sunlabob, partly through low-interest loans, and partly in kind by the villagers. One of the key advantages for the stakeholders was that they could in part privatise the existing infrastructure, which significantly lowered the costs. Helvetas also initiated a livelihood development project, RISE, based on the (assumption of) new opportunities created by the rehabilitated system. Most

Figure 4.6 Clockwise from the top-left corner: the forebay, the solar array, the powerhouse (Nam Ka 2), and the generator at Nam Ka

Source: Author.

of the labour involved in the project was provided by the villagers and failing to contribute labour was fined 20,000 LAK (US$2.5) per day. Therefore, many people were under the impression that they would be connected free of charge. It is unclear whether this confusion was a misunderstanding or deliberately left open during the many meetings.

In 2007, the rehabilitated hybrid system started operating. In order to get re-connected to the system, the villagers had to pay 900,000 LAK (US$112) per household. Furthermore, they had to start paying an electricity rate of 1,950 LAK/kWh (US$0.24) plus 10 per cent government tax with a planned increase of 8 per cent per year, or more than eight times the price of grid electricity in Laos.[8] Unlike the case of Phakeo, the tariff in Nam Ka was based on 'commercial rates'. However, the assumptions were curiously optimistic. The calculations assumed a connection of 200 households – while there were just over 100 in the village at that time – consuming an average of 2.4 KWh/day: a very high estimate for rural Laos.[9] The idea was that electricity would create productive usage and new livelihood opportunities – such as sawmills, rice mills, ice making – which would in turn create more demand. In practice, however, this proved to be much more difficult due to the high price of electricity, the technical limitations of the system, and the limited demand and skills to support this productive usage (Personal communication, 2012).

When the rehabilitation was finished, only 59 of the wealthier households (out of 101 in 2008) wanted – or could afford – to pay the connection fee for their own meter. Many of the people interviewed expressed discontent regarding the high costs of connection and use of electricity. Therefore, while not officially allowed, 31 households accessed their electricity through one of their neighbours or family members via a shared connection (RISE, 2008). The graph in Figure 4.7 shows the development of electricity consumption in Nam Ka since 2007. During the first few months, those who could afford it were happy to have access to electricity again, but they were not aware of the costs of usage. Many people who initially used a lot of energy and were shocked to receive a high bill a few months later. As a result, the village's energy use reduced dramatically, which can be seen in the second period from September 2007 until June 2010. The data suggest that over a fairly long period of time, electricity use remained stable at around 5 kWh/month, just a little more than the assumed usage over two days![10]

---

8   In 2011, for the first 25 kWh, the EdL tariff was 234 LAK/kWh (US$0.03).

9   To compare, the average electricity use in Sydney in 2011–2012 was 11.6 kWh/day (Ausgrid, 2012) and in the Lao residential sector 3.5 kWh/day in 2010 (EdL, 2011c).

10   It should be noted that this very low level of electricity use (even for Laos) was based on meter readings. By September 2009, there were at least eight meters broken, thus impacting on the average electricity usage (Personal communicaion, 2012).

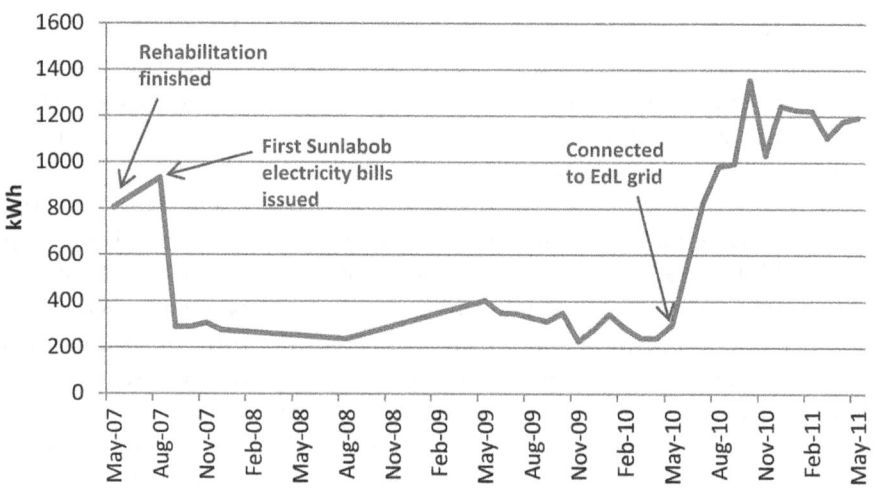

Figure 4.7   Total electricity consumption in Nam Ka per month in kWh from May 2007 to July 2011

Sources: Data from Sunlabob, Helvetas and EdL.

A few months after the system started operating, it became clear that the economic models used by Sunlabob were incorrect. The villagers did not nearly consume as much electricity as expected, and the maintenance costs proved much higher than expected due to technical problems. This meant that the company was running at a loss. Because of these financial problems, the diesel generator was never used, even when there was not enough water to produce electricity from hydropower and PV solar. In addition, because the villagers were very unhappy with the high costs of electricity, meters were tampered with and broken. Meanwhile, Helvetas was still waiting for permission from the Lao government to start their RISE project, which meant that they could not start supporting income generating activities. Because of the above factors, Sunlabob and Helvetas, supported by many people in the village, decided in mid-2008 to connect the village to the EdL grid instead and feed the electricity from the micro-hydropower system into the grid of this state-owned company. The implementation of this plan took more than a year due to prolonged and difficult negotiations between Sunlabob and EdL over an appropriate feed-in tariff, also because it was the first time this would happen for small-scale decentralised generation in Laos.

The construction of the grid, and the adaptation of the system, finally took place in the first half of 2010. During this time, there was no electricity available, despite Sunlabob's promise to provide an interim solution (Personal communication, 2012). By July of that year, the people of Nam Ka were able to get electricity from Electricité du Lao (EdL), and, by December 2010, all except 15 households had some form of electricity, either through their own meter or through an (illegal) shared connection. People without previous connection to the Sunlabob system had to pay EdL for the connection fee, at a minimum of 750,000 LAK (US$94) per household depending on the distance from the grid. By June 2011, all except three households were connected to the grid through a meter, 13 of them through a Helvetas scheme, providing interest free loans.[11] In terms of electricity usage, Figure 4.7 clearly shows that people changed their behaviour as a result of the sharp reduction in electricity tariffs. Interviews revealed that this increase was not only attributable to new users, but also to the purchase of new appliances and to increased usage of existing appliances. One of the deputy village leaders, for example, used to watch TV for one or two hours at night only. By the time he was connected to the EdL grid, he switched it on early in the morning and sometimes left it for most of the day. All of the people interviewed were positive about the reduced costs and the increased stability of their electricity system after connection to the grid.

Subsequent to the village being connected to the EdL grid, Sunlabob had to synchronise its micro-hydropower system so that it could sell electricity to EdL at the agreed price of US$0.06, half in US dollars and half in Lao kip (to minimise currency risk). Sunlabob also bought two new generators and installed them at the nearby 81 kW Nam Ka 2 micro-hydropower site. The company also synchronised that system to the grid, as the grid connection had made the system obsolete in the other villages too. Unlike the Nam Ka 1 system, this micro-hydropower generator – which was also funded by the Xinhua Company in 1995 – was still in working order at the time of grid connection in 2010, although one generator had failed and the voltage was unstable. Ever since the systems were synchronised, Sunlabob has experienced a lot of problems, significantly lowering the expected income from electricity sales. The company decided to stop trying to repair the Nam Ka 1 system and to focus only on the larger Nam Ka 2 system. However, in July 2011, both systems were damaged by a severe storm and landslides. They were still not functioning by September 2012 (Personal communication, 2012).

Reflections on Energy and Modernity

Both Nam Ka and Phakeo are cases of upland communities that have been increasingly confronted by state-led discourses and practices of modernity, through integration into the state and state integration into the village (Hirsch, 1989).

---

11   In other areas, poor people can get loans through a project called 'Power to the Poor' under similar terms. This project was initiated by the World Bank and administered by EdL.

The relatively long history of electricity projects in the village is one of the key examples of how outside projects embody the notion of development and modernity. Commenting upon the success of the projects in the village, a 22-year-old man said that the projects have brought some changes to the village, mainly in the forms of knowledge about farming equipment (through Helvetas) and ideas relevant to the organisation of housework. Although he was not sure of what the future would bring, he thought that the next generation would be 'more modern'. There was also the realisation amongst some people that as yet there is insufficient development: that more is possible. A 56-year-old man summed the situation in Nam Ka up as follows: even though the village now has a road, motor bikes and electricity, it is still not modern because the road is not good and many people are still poor. The last point was at the core of the plea of the village head, who mentioned several times that Laos is the poorest country in the world and that Nam Ka is one of the poorest villages in the country. He expressed the hope that people (like the author) would help to bring more development to the people of the village.

This case study clearly shows that the benefits of development projects do not affect everyone equally. One widowed mother, for example, thought it good that many families have been able to improve their livelihoods, albeit not hers. Other people observed that few of the development workers talked to them, let alone addressed their main concerns. The initial rehabilitation of the micro-hydropower system is another example, as only the richer half of the households was able to get the benefits of electric light and television; turning what was previously a common pool resource into a private good. In addition, while the standard of living had increased for most people, the costs of living had gone up too, creating new problems and dilemmas for people trying to manage their households. Again, the situation with the rehabilitated system is a case in point: villagers had to pay relatively large sums of money every month for a non-essential service. It was only after connection to the EdL grid that electricity became more affordable, although it still remained beyond the reach for the poorest in the community.

To summarise, the case of Nam Ka shows some of the issues that occur when energy trajectories are out of sync with the developments of peoples' livelihoods. While decentralised electricity generation is not a bad idea in itself, and electricity resonates strongly with the aspirations of peoples in rural areas to be 'modern', it failed in two ways in Nam Ka. First, the Chinese built micro-hydropower system could not be adequately maintained and eventually broke down. Second, the rehabilitation and livelihood improvement programme initiated by Sunlabob and Helvetas also largely failed due to the complex and expensive technology involved and the inability of villagers to afford the premium price for electricity. A qualification to these observations is the fact that grid expansion is heavily subsidised in Laos, creating an unequal playing field for decentralised generation (Martin and Susanto, 2014).

This concludes the two cases from Laos, one of the least-developed countries in Southeast Asia, and brings the chapter to two cases from Thailand, one of the wealthiest countries in the region.

Mae Kampong: Micro-Hydropower Eco-Tourism Showcase Community?

Mae Kampong, an upland village in Thailand, has been transformed from a remote upland community based on picking tea (*miang*) to a well-connected prosperous village with a booming tourist industry in less than 30 years. The micro-hydropower system, which was built in 1983, was one of the key drivers of – and catalyst for – these developments. Paradoxically, this 'success' has led to a decline in population and traditional livelihoods, mainly due to outmigration of the region's young people, like many other areas in higher-income countries in Southeast Asia. Moreover, since the village has been connected to the national grid, the importance and visibility of electricity and its infrastructure has gradually declined, increasing the disconnect between the place of production and the place of consumption.

Geography, History and Development Projects in Mae Kampong

Mae Kampong is a village in the sub-district Huay Kaew, Mae On district, Chiang Mai Province. Located approximately 45 kilometres east of the city of Chiang Mai, it is close to the border with Lampang Province (Figure 4.8). Both the provincial capital and its surrounding areas are popular tourist destinations, due to their mountains, waterfalls, hot springs and other tourist attractions. While Chiang Mai city is located in a valley, Mae Kampong is part of a hilly-cum-mountainous area, located at 1,300 metres above sea level and characterised by its forests and streams. The forests in the area are part of an officially protected zone, which was established to reduce the deforestation that has heavily affected Thailand (ICEM, 2003). The climate is relatively cool due to the high altitude.

The first settlements in the Mae Kampong area date back at least a hundred years (Wongsawat and Bhuntuvech, 2009), if not two hundred (Promjittiphong, 2005). Around that time, the ethnically Northern Thai people (*kon meuang*) started settling in the area, planting and picking *miang*, a type of tea. The Northern Thai people have their own history and language, both of which were historically linked to Laos but later heavily influenced by central Thailand. The *Lanna* (Northern Thai) kingdom had been a protectorate of Siam, before it

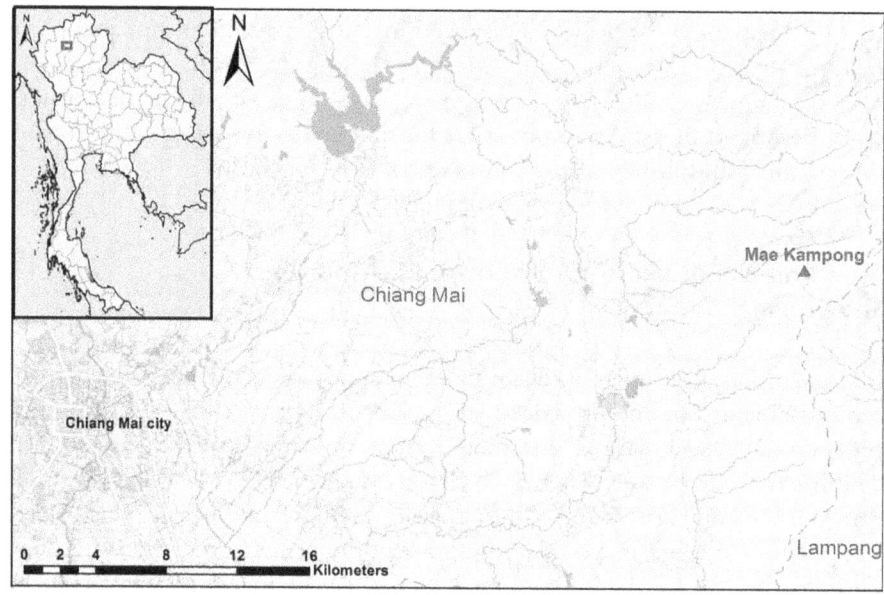

Figure 4.8   Map showing the location of Mae Kampong

Source: Author, adapted from Map data © 2015 Google.

was eventually fully incorporated in the Thai state in the early twentieth century (Winichakul, 1994).

During the Cold War and its aftermath, the rural areas of Northern Thailand were considered a threat to the US-aligned government because many alleged communists had sought refuge there. Huay Kaew, Mae Kampong's sub-district, was classified as a 'pink area' at that time.[12] This meant that the authorities were watching the activities in the area closely, particularly in the aftermath of the massacre of 1976 which ended three years of democratic government and reinstalled military rule. Several students who were on the run from the government hid in Mae Kampong and the surrounding villages during that time. The government saw the provision of infrastructure as a way to both control these areas and to get people on the side of the state. The construction of a gravel road (between 1973 and 1976) for example, reduced the travelling time from Mae Kampong to Chiang Mai city from about one day to 1.5 hours. The time was further reduced when the gravel road was upgraded to an asphalt road in 2000. In 1981, the first school was opened in the village. King Bhumibol Adulyadej paid a visit to the area and opened the Teen Tok Royal Project office,

---

12   Areas classified as 'red' were believed to be strongholds of communists and focal points of clashes between the communists and the government. Pink areas were considered a lesser threat.

with the objective of diversifying peoples' livelihoods by introducing new crops and new farming techniques.

The village of Mae Kampong was originally bigger than at present and stretched for more than six kilometres. Its Buddhist temple was built in 1930 and is still a central point of village life. In 1999, Mae Kampong was split into two villages, Mae Kampong and Tan Thong, because the village had become too stretched out and it was too far to walk from one end to the other. The current village comprises six hamlets, or clusters of houses, each a few hundred metres apart. In 2010, there were 132 houses and 367 people in the village, but this number was in decline due to the outmigration of the younger cohort. In 2005, for example, there were 417 people in 130 households (Promjittiphong, 2005, p. 22). With the village population declining and the price of land going up, there is increasing pressure on the locals to sell their land. Since around 2000, outsiders have bought land and a few resorts have now been established, usually by in-migrants but taken care of by local people.

Changing Livelihoods

In the past, people practiced swidden rice cultivation on the slopes of the mountains, but they stopped this activity some 40–50 years ago on the request of the government. Since then, the main livelihood activity in the village became picking and processing *miang*, which can be chewed after steaming and fermentation. Most of it is for export to other parts of the province. While this activity remains important for a number of people, the villagers' current livelihoods are more diverse and the incomes of most have increased. Many households now plant coffee in addition to *miang*: some have also started growing various kinds of fruits. Tourism became an important additional source of income from 2000 onwards. Table 4.4 shows some of the key indicators of the livelihoods of the surveyed people in Mae Kampong.

Among the surveyed group, approximately two-thirds of the households indicated that they were involved in the production and selling of *miang*, a tea plant that grows on hilly slopes at higher altitudes. There are three main seasons a year in which people go out almost every day to pick *miang* leaves. In the evening, the leaves are steamed, then packed in bunches to ferment prior to sale to traders. Chewing *miang* is popular amongst the older people, but the demand is declining. This has not affected the villagers' income to any great degree, because although production is much less than in the past and, over time, income sources have diversified. In the past, Mae Kampong attracted many seasonal workers from the lowland areas (outside of the rice production season) to help in the production and processing of *miang*. Today, this is no longer the case. Many villagers have stopped producing *miang* in the last two decades, instead producing coffee or doing non-agricultural jobs. As a result,

Table 4.4   Livelihood indicators of surveyed groups in Mae Kampong (n=18)

Category	Indicator	% of hh	Average	Minimum[1]	Maximum
Household size	Number of adults		2.4	1	5
	Number of children (<15 years old)		0.6	1	2
Agriculture	Land (ha)	89%	2.4	0.8	3.2
	Income from miang (THB/year)	67%	28,000 (US$935)	5,500 (US$183)	80,000 (US$2,667)
	Income from coffee (THB/year)	78%	13,300 (US$443)	3,000 (US$100)	40,000 (US$1,333)
Other	Income from homestay (THB/year)	28%	25,900 (US$863)	20,000 (US$667)	36,000 (US$1,200)
	Other sources of income	72%	58,100 (US$1,936)	10,800 (US$360)	228,000 (US$7,600)

[1]The minimum for all values higher than zero.
Source: Author.

some of the *miang* plantations are no longer picked: they are being 'given back to nature', to use the words of a former village leader. According to him, in 2011, approximately 10 per cent of the land which was in use before had already been abandoned.

Coffee has become an important cash crop for the people of Mae Kampong since its introduction in 1985. Initially, coffee trees were provided free of charge by the Royal Project office, but now people invest by themselves. The crop, which is well suited for the terrain and the climate, is often intercropped with *miang* trees, by extension requiring little adaptation to the older agricultural practices. Many people switched from *miang* to coffee because the latter requires less work and can yield higher profits. At first, all of the coffee was processed outside of the village, but later the village head acquired a small coffee roasting machine after which (part of) the produce was roasted locally. Where most of the coffee is sold in bulk to local and international companies, part of it is retained and sold to tourists visiting the village.

Around 2000, tourism became another important aspect of the Mae Kampong villagers' livelihoods. One of the first ventures was the introduction of a homestay programme, which has become very popular and proven profitable for the village. This programme lets people sleep and eat for one or more days in one of the rooms or house extensions of a family for 350–550 THB (US$11.70–18.30) per person per night, depending upon the length of stay. In 2011, 21 households were certified to do homestay, and all of the revenue goes through the village cooperative. The cooperative, in turn, enforces the standard of the facilities, for example private rooms, a large and clean bathroom, and hot showers. Another noteworthy tourism development is the Flight of the Gibbon project, a form of eco-tourism providing a 'zipline adventure experience' through the forest just outside of the village. The project, which was set up in 2008 with three kilometres of ziplines, was expanded in 2011. At the time of fieldwork in 2011, this venture attracted more than hundred people every day. The land required was rented from people in the village: the project paid 100,000 THB (US$3300) in the first year, 20,000 THB (US$670) in the following years, and 100,000 THB every fifth year. Part of all these rents goes to the cooperative.

Both homestay and the Flight of the Gibbon have created employment and income generating opportunities for the villagers. They include preparing food, running coffee shops, providing massages, making stuffed *miang* pillows, playing music, doing cultural performances, and guiding tourists around the village. Many of these opportunities have been taken up by older people rather than the adolescents in Mae Kampong. In addition, most of the employees of the Flight of the Gibbon come from outside the village, because they need to speak some English. Other occupations according to those interviewed included dog breeding, the making of various kinds of handicrafts, selling orchids, construction work, and government jobs.

Energy Trajectory

The context of livelihoods aid to understand the embeddedness of Mae Kampong's energy trajectory, which is shown in Figure 4.9. Before 1982, people in Mae Kampong had virtually no access to electricity. They would use kerosene lights and gas lamps to provide light in the evening, mainly to steam and package *miang*. A few people had diesel generators and some used car batteries for watching TV and providing light for a maximum of a few hours per day.

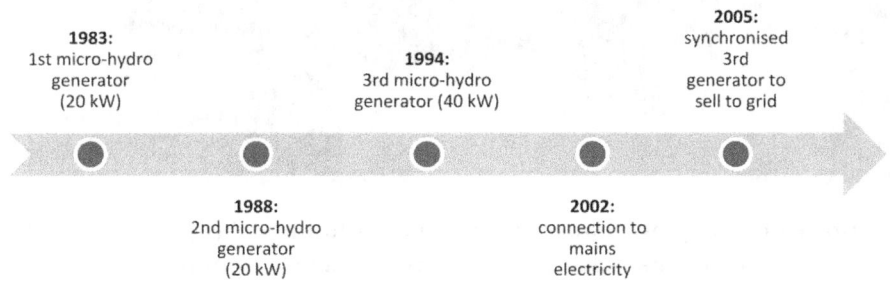

Figure 4.9   Selected events in Mae Kampong's energy trajectory

Source: Author.

In 1982, construction started on the first micro-hydropower generator, one year after King Bhumibol visited the area (Figure 4.10). The Teen Tok Royal Project office initiated the project and funding was provided by USAID. Technical support came from the Department of Alternative Energy Development and Efficiency (DEDE), which also helped to set up a cooperative to carry out maintenance, collect money, and manage the revenues from the hydropower system. The villagers were involved in the construction of the 50 metre penstock, the powerhouse, and in setting up the village grid. The capacity of the first generator was 20 kW: it served approximately 150 households both in Mae Kampong and in part of (what is now) Tan Thong village. Soon after the installation, people started buying TVs, fridges and other electric appliances, leading to rapid growth of electricity demand (see Figure 4.11). Initially, the committee tried to limit the amount of electricity usage of each household to three light bulbs and two power points, but this measure proved difficult to enforce.

With the financial support of DEDE, a second 20 kW micro-hydro turbine was added to the existing powerhouse in 1988 in response to increased demand and to electrify Mae Laai, another small neighbouring village. This village bought electricity from Mae Kampong's micro-hydropower cooperative, enabling the latter to make a significant profit between 1984 and 2000 and to expand activities to other areas such as social security (hospitals, cremation), loans (since 1993), and tourism (from around 2000). In 1998, due to the increasing number of

Figure 4.10 The powerhouse (left), generators 1 and 2 (upper-right) and two penstocks (lower-right) in Mae Kampong

Source: Author.

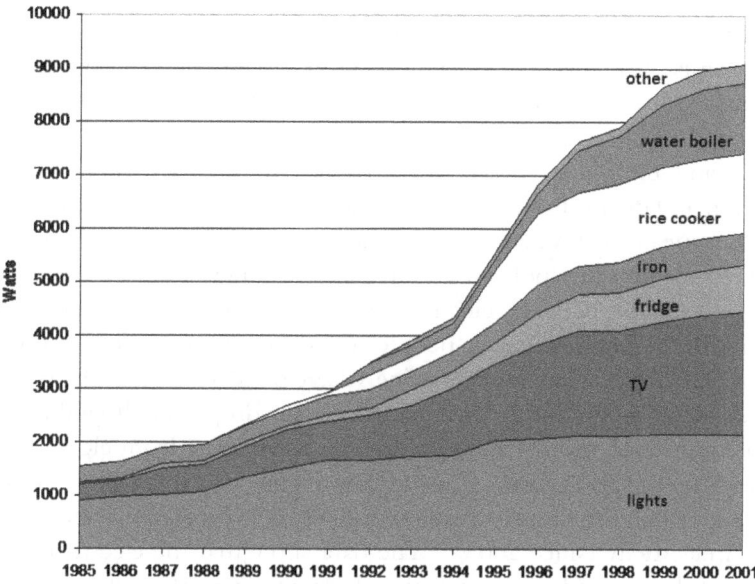

**Figure 4.11 Growth in peak electricity demand in Mae Kampong,
broken down by appliance, 1985–2001**

Source: Greacen (2002), reproduced with permission.

appliances and natural growth in the three villages of Mae Kampong, Tan Thong and Mae Laai, a third generator (40 kW) was installed further downstream.

As early as 1993, talks commenced about connecting Mae Kampong and its neighbouring villages to the national grid. This led to some conflict within the village, because certain groups wanted to keep their own micro-hydropower system. There are different narratives about this particular episode, which provide insight into the framing of the issue, and into the conflict that erupted over which type of energy-modernity to choose in Mae Kampong. According to the current head of the cooperative, there was no conflict, merely some misunderstanding because the Provincial Electricity Authority (PEA) was slow in constructing the electricity poles and lines. In addition, he said, the people of Mae Kampong were afraid of the new high voltage lines running through their village. The head of the sub-district at that time said it was mainly an economic issue: the cooperative in Mae Kampong did not want to forfeit the micro-hydropower revenues on which it was dependent. The current village leader of Mae Kampong, who held the position of treasurer of the cooperative at that time, agreed that loss of income was indeed the issue which divided the village into two factions. Those who wanted to keep the extant micro-hydropower system were accused of 'opposing modernity' by those in favour of grid connection.

After long negotiations over the extension of the grid to Mae Kampong, a compromise was reached. Every village in the sub-district with an existing decentralised electricity supply would be able keep their own decentralised system in parallel if so desired. The grid extension to Mae Kampong was completed in 2002, and the village cooperative decided to keep its micro-hydropower generation and distribution system in parallel to the new distribution lines from PEA. This meant that Mae Kampong had two electricity grids, with some parts connected to the same poles. Each house had either two separate circuits, or a switch to choose between electricity from their own micro-hydro system or from the PEA. In some cases, people had different appliances connected to the different circuits. The neighbouring village of Mae Laai, on the other hand, made a collective decision to stop using micro-hydropower altogether in 2004.

The amount of money collected by the cooperative from electricity slowly declined after the grid arrived. The graph in Figure 4.12 shows the development of electricity use for the two systems for micro-hydropower (since 2003) and for PEA electricity (since 2008).[13] The data show that the electricity from the national grid was used more than the electricity from the village hydropower

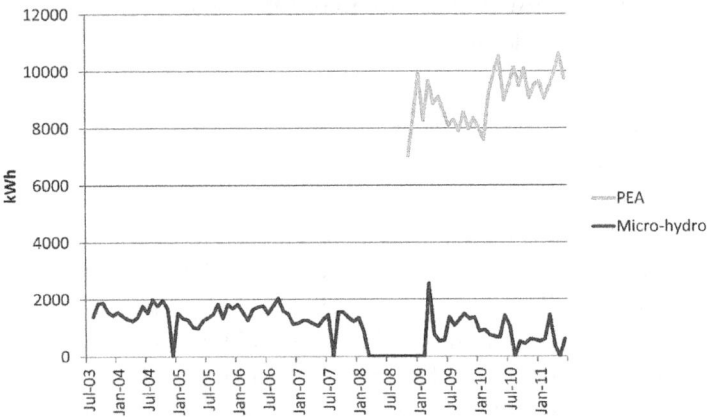

Figure 4.12 Electricity consumption of PEA and micro-hydropower in Mae Kampong

Sources: Author, data from Mae Kampong micro-hydropower cooperative and PEA.

---

13   Two types of data were used to create this graph, the first based on the money collected by the micro-hydropower cooperative. To convert this data into kWh, it is divided by the electricity price, 2 THB/kWh (US$0.067). One of the complications is that the base payment was 10 THB (US$0.33) per household per month for the first 5 kWh, regardless of usage. The calculations assume that 50 people paid 10 THB (US$0.33) per month, but did not use the electricity (as indicated in the interviews). The data from PEA are more straightforward, but only available since January 2009.

generators. Moreover, while the use of electricity from micro-hydropower gradually decreased between 2003 and 2011, the (limited) data on the use of grid electricity suggest an increase over a two and half years' period. The reasons for this decline in micro-hydropower usage were due to less consumption and some users opting to disconnect altogether. Between 2006 and September 2010 alone, 43 households stopped using micro-hydropower in Mae Kampong and Tan Thong, the majority in the latter. By 2011, only 21 houses in Tan Thong village still used hydropower compared to 113 in Mae Kampong. Furthermore, many people chose to use electricity from PEA rather than micro-hydropower electricity. When the government started to provide free electricity to people using less than 90 kWh per month in 2006, many villagers stopped using micro-hydropower altogether. However, some retained their connection and payment of 10 THB (US$0.33) per month, simply in order to retain the benefits that came with the membership of the cooperative.

The decreasing importance of the micro-hydropower system shows clearly in the costs and revenues of the cooperative's annual report of the cooperative. In 2001, income from hydropower accounted for 44 per cent and homestay 34 per cent of the total income of the village cooperative. In the following year, the income from homestay had more than doubled, from 87,000 to 173,000 THB (US$2,900 to 5,770), accounting for 58 per cent of total income. During the same period, income from hydropower declined in both percentage terms (to 24 per cent) and absolute figures, from 113,000 to 71,000 THB (US$3,770 to 2,370). By 2009, the income from the homestay program was a staggering 1,511,000 THB (US$50,400), while the income from hydropower was only 30,000 THB (US$1,000). Furthermore, since the connection to the PEA grid, the micro-hydropower system expenses often exceeded its income (as shown in Figure 4.13). And, because the expenses did not include depreciation and major repairs (such as the replacement of poles and wires in 2007 at 82,500 THB (US$2,740)), the village is now effectively subsidising its two micro-hydropower generators through income from tourism.

Another development in the energy trajectory was the synchronisation of the third (40 kW) generator to the grid, to start selling electricity under the Thai VSPP regulation. This process, which was carried out with support from DEDE and PEA, took approximately 6 months, after which it generated over 35,000 THB (US$1,170) per month in income. After less than a year, however, electricity sales were stopped because it became clear that it was against the law for one government body, in this case DEDE, to sell electricity to another government body (PEA). As a result, no electricity has been sold since early 2009: the third generator is effectively useless as it can no longer provide electricity to the village. This has ostensibly left negotiations in a deadlock. Mae Kampong does not want to share its revenues with the other villages in the sub-district as proposed by the sub-district administrative organisation (SAO).

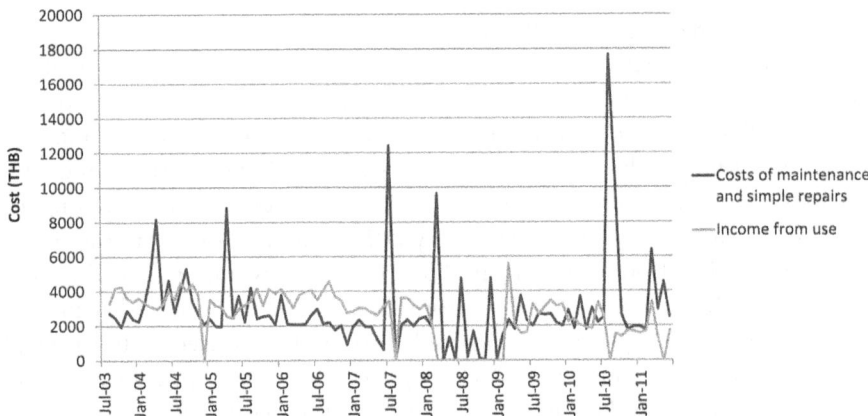

Figure 4.13 Income from use and expenses of maintenance and simple repairs for the first and second generators in Mae Kampong, 2003–2011

Source: Author, data collected by author from micro-hydropower cooperative logbooks.

The preferred option of many in Mae Kampong villagers of reverting to the original system – using the third generator as stand-alone again – would cost approximately 0.5 million THB (US$16,700).

Reflections on Energy and Modernity

Mae Kampong is yet another example of an upland community going through a series of energy transitions that have profoundly shaped the livelihoods and types of development in the village. In the space of three decades, the village has been transformed from a remote community dependent upon *miang* for survival into a well-connected and economically prosperous village attracting increasing numbers of tourists. However, there are a number of new challenges and downsides to the energy-modernity in Mae Kampong, such as threats to the environment and liveability of the area. For example, one deputy village head observed that garbage collection has become more difficult because of the many tourists. While some people still burn their waste like they used to, there are efforts to separate and sell it and collect and dispose of plastic bags. Another development impact mentioned by some is the increased level of noise. A 35-year-old man, who married into the village in 2008, said that while the recently completed road to Lampang is convenient, it has led to problems of noise in the village.

Mae Kampong is an important illustration of the paradox of development and modernity in rural Thailand, where increasing prosperity and improved quality of infrastructure appear to have dissuaded some people from leaving the countryside. Conversely, in many ways, modernity in Mae Kampong has contributed to the outflow of the youthful cohort, who can now access better education outside of the village and often do not want to return. The latter effect seems to prevail and Mae Kampong, like many other villages in Thailand, now has an aging and shrinking population, which is increasingly dependent on tourism and remittances. Some of the *miang* fields were already being abandoned and many more would follow. Even the eco-tourism success has not been able to create an incentive for people to stay in the village. In addition, there is increasing pressure on the local people to sell their houses and land to outsiders for holiday homes, leading to increasing prices of land.

There is, however, increasing awareness in the village of the downsides of increased simple modernity beyond issues such as noise pollution or waste. One of the representatives of the Sub-district Administrative Organisation (SAO) said that once you start development, you cannot stop. You need to solve the problems and move on, because people want to generate more income. Moreover, he suggested that development should take the preservation of the environment and need for conservation into account. Others, for example the 43-year-old deputy village head, were more explicit. He considered that there is enough 'development' in the village already, and that people should now focus on maintaining the existing systems, such as their water supply, electricity system and the forest. He also observed that being 'too modern' was not good for the village.

Finally, this case study shows how electricity has moved gradually from something very important and visible in the village to something that has become less and less important and nearly invisible. When the micro-hydropower system was constructed, it became possible for everyone to enjoy the benefits of light, television and other appliances. As such, electricity was an important way in which people could access the world beyond the village. Over time, however, electricity became slowly taken for granted and the production of it less important to the local economy, even when the actual demand and consumption of electricity kept increasing. Moreover, people increasingly demanded stable 24-hour electricity as they became less tolerant of black and brownouts, increasing the disconnect between the place of production and consumption.

Bo Nok: From Coal-Fired Power Station to Renewable Energy Experiments

The energy trajectory of Bo Nok is different compared to the other three case studies, because Bo Nok is known in Thailand as the sub-district that was targeted as a site for a coal-fired power plant. However, the plant never materialised due to the resistance from within and outside the community. Over time, Bo Nok has become associated with resistance against large state-led discourses of modernity and projects which has opened spaces for experiments with decentralised electricity generation.

Geography, History and Changing Livelihoods

Bo Nok is a sub-district of Prachuab Khiri Khan District in Prachuab Khiri Khan Province, which connects Central and Southern Thailand through long and small strips of land in which some places are only about 11 kilometres wide. Bo Nok sub-district itself is situated approximately 25 kilometres from the provincial capital, Prachuab Khiri Khan City, and less than 10 kilometres from the district capital of Kui Buri. In the east, it connects to the Gulf of Thailand, while in the hilly west it connects to Myanmar (see Figure 4.14). The sub-district includes 14 villages, which are nowadays somewhat arbitrarily divided by roads, canals, and the railway line. The climate in the province is characterised as monsoon-tropical, with temperatures between 10 and 40 °C, with an average of 27 °C. The weather is usually dry and hot and therefore suitable for arid crops such as pineapples and aloe vera. The beaches in the east of Prachuab Province are an increasingly popular tourist destination.

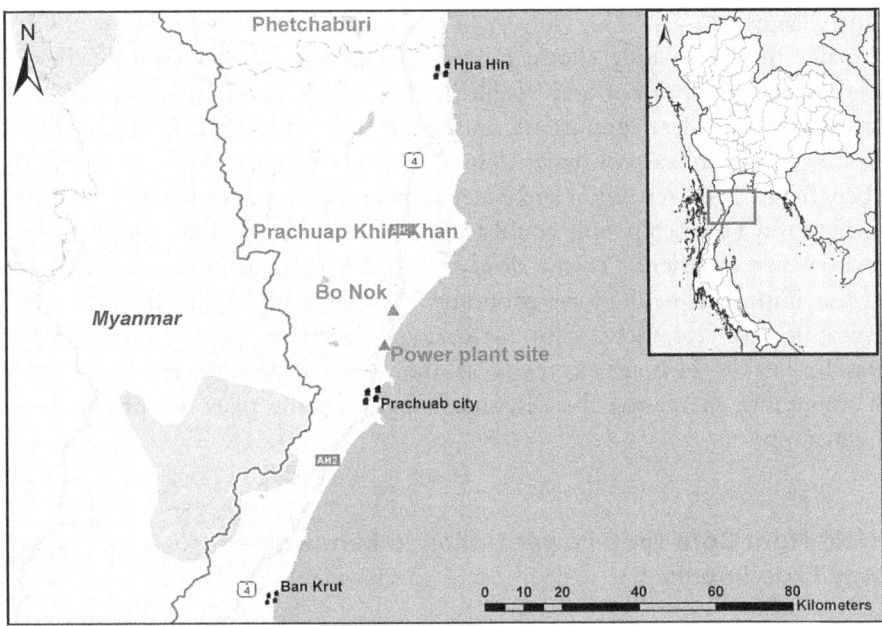

Figure 4.14 Map showing the location of Bo Nok sub-district
Source: Author, adapted from Map data © 2015 Google.

There are documented settlements in the district as early as the Ayutthaya era (1350–1767). However, most of the people, especially the elder, remember the time when the sub-district was still mainly covered in forest with a few small settlements, most of which were located near the sea and dependent upon fishing. Besides fishing, people used to plant rice and have small plots in which they grew other agricultural products, such as bamboo and chilies. The current livelihoods in the province, by comparison, are very different and highly diverse.[14] Today, there are hardly any remaining rice fields, neither in the sub-district, nor in the rest of the province. The provincial slogan gives some indication of the current livelihoods: *City of pure gold, delicious coconuts and pineapples, delightful beaches, mountains and caves, land of spiritual beauty.* Many people now grow cash crops, such as pineapples, coconuts, aloe vera and mangoes. According to the provincial office, pineapples and coconut were the two key agricultural commodities in the province in 2011, as well as in the sub-district of Bo Nok (Prachuab provincial office, 2011). Oil palm plantations have become increasingly common as well over the last few years.

The mode of production in agriculture has changed from smallholder-based to capital-based agriculture. In the past, people grew their crops on small plots of land, and ownership passed down through inheritance, like in the two cases in Laos. While some families still operate in this way, large agricultural plots owned by wealthy families inside or outside the district are very common now. People with fewer assets rent plots of land or work as wage labourers on these larger estates. Some villagers also claimed that agriculture had become more difficult. A 65-year-old woman, for example, said that it is difficult to invest in agriculture these days, because of the dry weather, limited access to water and poor soil conditions. She added that in the past, many people could grow rice, but this is not the case anymore. Moreover, the land suffers from salt intrusion, which makes it unsuitable for many types of crops. There is also some livestock (mainly beef cattle, pigs and chicken) in the sub-district and the province.

Fisheries are another important source of income, especially for villages located close to the sea. It was not only an important livelihood strategy for the people of Bo Nok, but also for migrant workers from Myanmar and Cambodia. Many families still use small fishing boats, but the larger boats now in use make it harder for small fishermen to make a living. A 22-year-old fisherman said that fishing has become more difficult in Bo Nok because there are more boats and people using more advanced technologies, such as pulling nets. The locals have to go further away, sometimes to other provinces. Moreover, the number of migrant workers has decreased as many among them now go to Bangkok for

---

14 Because of this diversity, there is no livelihoods table (unlike the previous three case studies).

work. In order to generate more income, this fisherman's family started to grow potatoes in 2011, seeing it as a way to deal with the difficulties of finding fish.

Shrimp farms are also an important feature of the landscape around Bo Nok. However, they are a high-risk investment. There are many stories of people who tried to grow shrimp, but failed to make a profit. Incidences of disease, as well as high volatility of price, have led many shrimp farms into bankruptcy. These days, the owners of shrimp farms are usually not smallholders, but people who own several shrimp farms in the area. The daily work on the farms is done mainly by migrants, while the sorting is usually left to local female day labourers.

Work in the industry and services sectors has become much more important in the last few decades. In 2011, there were 612 factories throughout the province, employing a total of approximately 22,000 people (Prachuab provincial office, 2011). Industry also accounts for the largest share of the Gross Provincial Product (22.3 per cent), followed by trade (13.5 per cent) and agriculture (13.1 per cent). In Bo Nok sub-district, the main form of industry is a large fruit factory on the Southern Highway, which produces canned fruits and juices, mainly based on pineapple. Besides factories, there is also employment in government offices, shops, markets, and other forms of services (Prachuab provincial office, 2011).

Finally, tourism has proven an important way for many people to generate income. Hua Hin is the best known tourist destination in the province, but other places along the coast, such as Bo Nok, are becoming increasingly popular. There were already several holiday resorts and more under construction in 2011. Another way tourism manifests itself in the area is through holiday houses. Most of the people who lived near the coast sold their land to developers, who target wealthy foreigners for their new beach villas.

Energy Trajectory

The energy trajectory of Bo Nok differs from that of the other three case studies, given that the sub-district has been connected to grid electricity for many decades. Initially the grid covered the main roads, but grid expansion has ensured that nowadays almost everyone has access to this type of electricity. The rate of development of grid electrification for the households interviewed appears in Figure 4.15, which shows that the first interviewee gained access to electricity around 1970 after which the sub-district was rapidly electrified. By the 1980s, 80 per cent of all interviewees had access to grid electricity and all of the surveyed households had access by 2001. Throughout the province, 97.3 per cent of all households had access to grid electricity by 2007 (Prachuab provincial office, 2011). Bo Nok's energy trajectory, however, is also marked by the controversy and conflict over the siting of the coal-fired power plant and its aftermath.

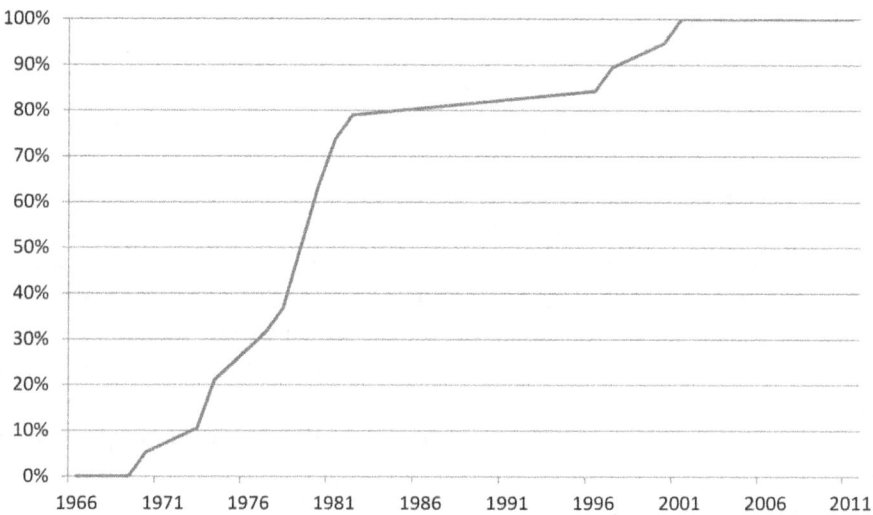

**Figure 4.15 Grid connection in Bo Nok amongst surveyed households
over time (n=19)**

Source: Author.

In 1995, some people in Bo Nok noticed that large amounts of land were being sold. They were told that the site was designated to be a golf course and resort. But, as they found out later, the real purpose was to build a coal-fired power plant (by the Thai-American Gulf Company). By that time, the government had already approved the plans and, in the next two years, the company was granted permission to build the power plant. It accordingly signed the Power Purchase Agreement (PPA) for a 734 MW power plant, to be extended to 1,400 MW at a later stage. This power plant, and another in the same province (Hin Krut, 1,400 MW), were part of a first call for IPPs in the context of liberalisation of the electricity sector in Thailand.[15] Both the position of Prachuab Khiri Khan province in a 'voltage dip' between Bangkok and Southern Thailand, and emerging plans for a large industrial zone on the Western sea board, were the key reasons for targeting this province.

Soon after the plans became public, the community – under the leadership of Charoen Wataksorn – mobilised itself against the development of the power plant, through legal and other support from academics, politicians, local and international NGOs, agencies such as TERRA, Alternative Energy Project for Sustainability, Palang Thai, Greenpeace, and Probe International (Kuze, 2003). The conflict, which became one of the key environmental controversies in

---

15 A third non-IPP 1,400 MW power plant in the province, planned for Thap Sakae and owned by EGAT project, was also on the table until 1999.

Thailand, lasted for about seven years. Table 4.5 summarises the key moments in the conflict.

The local conservation group's key objections, as summarised by Kuze (2003), were related to heavy air pollution, discharge of cooling water into the sea, the construction of a 3.5 km-long pier, waste transportation and repository, lack of community participation in the decision-making process, redundancy of the power plant, and a PPA which strongly favoured developers. In order to address these issues, the above group organised itself and employed several strategies to try to get the attention of the government. In Bo Nok alone, 94 letters were sent, 28 protest rallies organised, and five negotiation processes entered into. Additional seminars, rallies, letters, and negotiations were organised in cooperation with the movement in the other targeted area in the province, Hin Krut (Kuze, 2003). A community radio was set up to spread both information and knowledge and call for meetings and rallies. Perhaps the most significant of all activities involved the blocking off of the Southern Highway at the end of 1998, which attracted nation-wide attention.

Besides the large group of people who were against the plan to build a power plant, there were also many people in favour, amongst which were some local village leaders and politicians. Although it was difficult to get clear figures on the number and geography of the people on either side of the conflict at the time, there are some indications. Some villages in the sub-district were known to be in favour or against the power plant, and most of them still maintain their position. For example, villages dependent on fisheries and tourism tended to be against the project, because of the more immediate impact on their livelihoods. In the sample for this research, 5 per cent of the people were in favour of the power plant, 33 per cent said they were neutral and 62 per cent were against. Moreover, just under half of all people interviewed said they actively participated in rallies either in favour or against. This was similar to the findings of Kuze (2002), who compared the attitudes of people in agricultural production groups in different villages.

For many years, the situation in Bo Nok was effectively in deadlock. The local conservation group tried to find ways to stop the project and the project developers tried to win the hearts and minds of the people using various means. Thaksin Shinawatra's election victory in 2001 eventually led to a turning point. However, it was not until January 2002 that Thaksin finally visited the area. At least 10,000 people in Bo Nok (and 4,000 in Hin Krut) showed up to rally against the power plant. Finally, in May 2002, the government decided to freeze the projects for an unspecified amount of time. A few years later, the IPP contracts were changed into gas-based electricity generation and the plants were built in the nearby provinces Saraburi (by the Gulf company) and Ratchaburi (by the company active in Hin Krut) (Kuze, 2003). In 2011, the land was still

Table 4.5   Activities by government, the Gulf Company and the local conservation group relevant to the power plants in Bo Nok and Hin Krut

Year	Date	Actions of the government and the Gulf company	Actions of the Bo Nok local conservation group
1995	20/02	The company clarifies the plans to villagers in Bo Nok.	The people in Bo Nok show their first objection.
1996	30/11	Gulf company obtains permission from EGAT for IPP project.	
1997	25/07	Bo Nok Sub-district Administrative Organisation (SAO) submits people's protest letter to provincial governor.	More than 1000 people in Bo Nok submit a letter to Bo Nok SAO and gather in front of the provincial office. They also block the road.
	20/10		More than 2000 people from Bo Nok and Hin Krut gather together for the first time against the power plants.
	22/12	Gulf company signs PPA with EGAT for Bo Nok power plant.	
1998	8/12 (until 10/12)		10,000 people from the local groups gather in front of the provincial office and close the Southern Highway at Bo Nok junction.
	15/12	Cabinet agrees to organise a public hearing about both power plants (Bo Nok and Hin Krut).	
1999	09/03		Bo Nok and Ban Krut groups submit an open letter to object against the public hearing process.
	21/12		Bo Nok SAO submits a letter to the cabinet voicing their disapproval of the project and asking for the outcome of the public hearing.
	28/12		Around 800–1000 people from the Bo Nok group submit their demands to the Prime Minister at the parliament.
2001	28/02		Bo Nok local conservation group submits a letter to Prime Minister Thaksin.
2002	24/01	Prime Minister Thaksin goes to Bo Nok.	
	10/05	Thaksin announces the postponement of the projects for two years.	

Source: Translated and abridged from Kuze (2002).

owned by the company, and some people thought that the company and the government still wanted to use it to build a power plant.

The aftermath of the conflict involved a series of legal cases surrounding the acquisition of land and the corruption involved in the power plant planning process, which led to the assassination of the leader of the movement, Charoen Wataksorn. On 21 June 2004, he was shot by two people, near the temple *Wat See Yek Bo Nok*, immediately after testifying in a court case about land grabbing related to the power plant conflict. At the time of fieldwork in 2011, the local conservation groups of Bo Nok, Hin Krut and other places in Prachuab were still very active and often met to discuss issues of local development and environmental concerns such as fisheries, industry, waste management and national parks. There were at least seven active core groups in four districts in the province, which met both individually and occasionally jointly.

The aftermath of the conflict saw the development of several renewable energy projects, initiated by the government, the community and the private sector. One of the community-led projects aimed to build three locally designed windmills to generate electricity. Two of the windmills were installed on the land near the temple that supported the movement, *Wat See Yek Bo Nok*. These windmills are connected to a set of batteries which partly powered the equipment for the community radio. A third one was set up on the beach near the movement leader's bungalow park. Another example was the installation of solar panels on the roof of a school in village 9 of Bo Nok sub-district. In 2004, the roof of the school was covered with 16 solar panels (with a total of 1,825 Wp) provided by Greenpeace, to whom they were donated as well. The electricity generated was to be used in the school or fed into the grid (using a two-way meter). However, the system worked only for two or three years and it has been down since. Despite repeated attempts by a teacher, Greenpeace had not been able to get it fixed. A final example is the efforts of some people to plant approximately 5 hectares of palm oil on the temple grounds of *Wat See Yek*, one of the main gathering points during the conflict. They were hoping to be able to sell the oil at the community gas station, and to earn around 50,000 THB (US$1,670) per month from this venture.

There is also a prominent example of a private sector-led renewable energy initiative in Bo Nok. A company called Green Energy Technology built a 1 MW solar farm in accordance with the VSPP legislation, and completed it at the end of 2011 (Figure 4.16). According to the director of the company, PEA had specifically suggested Bo Nok sub-district as a site for their first project because of its turbulent and violent history over the siting of the coal-fired power station. They employed a novel concept based on producing electricity while at the same time avoiding the loss of employment and agricultural land. They did so by mounting solar panels about four metres high and combining them with crops underneath and (optionally) hydroponics in between, a

Figure 4.16 The 1 MW solar farm in Bo Nok (left) and jatropha plants under the panels (right)

Source: Author.

technique unique to Thailand. Despite the link to the conflict, the level of engagement and knowledge in the community about this solar project seemed low during interviews.

The government of Prachuab also proposed several projects to implement alternative energy, as explained by the director of the provincial energy office. Some of these projects are explicitly linked to the power plant controversy. A 5 MW solar farm was planned to be established near the Hin Krut power plant site, which would cost approximately 625 million Baht (US$20.8M). The 32 ha land required had already been bought, but the local conservation groups protested against the plan, advocating that the vacant land originally designated for the cancelled power plant should be used first. Another project, a 9.4 MW biomass-based power plant, was also targeted by the campaigns of conservation groups in the province, who feared that the feedstock would be complemented with coal and were critical about the waste management plan. In addition to the above, there were also several smaller, and less contested, government projects. One of them was a community energy project, which involved around 100 sub-districts, which stood to gain 200,000 THB (US$6,700) each for household level biogas systems. The sub-districts which were involved in the conflict surrounding the construction of the power plants were to be considered first for inclusion. By 2010, 40 such systems had been constructed; furthermore, between two and three thousand improved cook stoves were distributed among houses in each sub-district.

Reflections on Energy and Modernity

The conflict over coal-fired power plants in Bo Nok's recent past made people very articulate when expressing their ideas about development and modernity. While superficially the conflict that gripped Bo Nok was about the construction of a coal-fired power plant, in essence it touched directly on some of the tensions between modernity and sustainability, in particular in relation to the role of the state and its discourses of energy-modernity. Some framed the desired type of development as a choice between 'traditional' agriculture and 'modern' industry. A 40-year-old man, for example, said that it is better for people to work in agriculture rather than in factories, because the former is the traditional occupation of the people in the area. The current leader of the protest movement in Bo Nok echoed this comment when she said that the people were small-holder farmers and fishermen, but that the government did not appreciate these kinds of activities because they do not show in the GDP figures. Opponents, however, stressed the need for more industrialisation. One of the former village heads said that if the power plant were to be built, it could attract industries with high power demand and generate income and employment for the sub-district. As things stand, the sub-district was left with no development, no income, and with large agricultural debts. A former head of the sub-district also stressed that the main challenge for development was the creation of employment. In this respect, the big factories and other industrial developments could prevent people from having to migrate to the city in search of work. Another 59-year-old male supporter of the power plant suggested that Bo Nok had missed the chance to follow the models of districts like Hua Hin and Bang Sapan Yai, which have developed due to the tourism and the iron smelting industry respectively. Finally, there were those people who did not see agriculture and industry as necessarily opposed. The abbot of Wat Bo Nok, for example, thought it would be possible to have industrial development in addition to agriculture if it was done properly.

For some people, the ultimate question was whether or not to be 'developed' or 'prosperous' (*charoen*). A 35-year-old woman and a 73-year-old man observed that if the power plant was built, the sub-district would be more developed. A 54-year-old woman said that other places that had power plants were 'more developed' than the people in Bo Nok. Interestingly, all of these people claimed to have been 'neutral' in the conflict. Another 'neutral' 58-year-old man opined that even without the power plant, the people of Bo Nok could still be prosperous. Some people did not even see the connection between the power plant and development, such as a 28-year-old female who thought that building a power plant would not change much because people would still have to work in agriculture.

Like in Mae Kampong, some villagers thought more critically about the issue of development and modernity. One active member of the local conservation group, for example, thought there was too much development already: the province had a lot of natural resources and could grow many different products. In addition, he said, for those working in agriculture, economic crises do not really matter, because people need to eat anyway. The abbot of *Wat See Yek Bo*

Nok observed that in the current society, people are encouraged to have more and more things – for example to have luxury lifestyles, get used to the feel of air-conditioning – all of which favour large-scale shopping malls over the local market.

Finally, the case of Bo Nok shows how the plans for the development of a coal-fired power station have positioned this sub-district at the centre of negotiations about different discourses of energy-modernity in Thailand, which are echoed in many other parts of Southeast Asia. Following contestation of this state-led modernisation project, space has opened up for experimentation with alternative energy-modernities based on small-scale or larger scale use of renewable energy. This experimentation has had influence well beyond the sub-district itself, leading in effect to the re-framing of energy-modernity at the national level, discussed in the previous chapter.

The Context and Embeddedness Energy Trajectories

This chapter has presented four case studies in Laos and Thailand as energy trajectories embedded in processes of livelihood change at the local level. In each of the four cases, energy systems changed in accordance to changing discourses of development, with notable shifts in discourses around systems of energy provisioning. Moreover, each of the cases show how changing discourses of energy-modernity at other scales – such as the national scale analysed in the previous chapter – directly or indirectly influence peoples' livelihood and energy use patterns, as well as energy governance at the local scale. At the same time, the local social and political changes analysed here also influence what happens on these broader scales, in particular in the case of Bo Nok but also for other cases. While the successes or failures of each of the energy systems are open for debate, the energy trajectories illustrate that one cannot understand energy trajectories outside of the context in which they unfold. To this end, this chapter has set the scene for the next chapter, which shows how both cases relate to processes of modernity in everyday life.

Chapter 5
Energy and Modernity in Everyday Life

Introduction and Key Arguments

This chapter explores the everyday dimensions of transitions in household end-use technologies in Southeast Asia, drawing on empirical data from the four case studies introduced in the previous chapter. The aim is to show how energy trajectories are related to assets and equity of households, the uptake of different technologies, and what these changes mean for the everyday lives and experiences of modernity in the four case studies. Thus, while Chapter 3 adopted a largely national perspective, and Chapter 4 a village perspective, this chapter focuses on villages, households and individuals. Moreover, in contrast to the previous chapter, which discussed the case studies in sequence, this chapter will treat the case studies in parallel to draw out the different issues and dimensions. Whereas there are some comparative elements in the four case studies, the main aim is to highlight the different aspects of energy transitions in everyday life rather than simply compare the four cases. Importantly, this chapter identifies trends and demonstrates contingency rather than providing exact answers. Therefore, many of the processes and arguments carry beyond the borders of the case studies and are relevant to other parts of Southeast Asia.

The first key argument of this chapter is that the influence of energy trajectories on the local scale cannot necessarily be explained in terms of meta-processes and trends (Rigg, 2007). In order to substantiate this argument, this chapter links energy-modernity with appliances use, culture, travel behaviour, and changing worldviews. These linkages vary for different groups of people, such as the young and the old, the rich and the poor, or between different ethnic groups.

Second, this chapter argues that energy transitions are related to the changing dimensions of everyday modernity, such as different experiences of time and distance, the shift in balance from community to the individual, and the increasing 'invisibility' of energy systems in everyday life in the four case studies (Shove, 1997).

Third, the chapter shows that the influence of the energy trajectories on environmental sustainability is just one of the dimensions of changing energy-modernities. Moreover, this influence is hard to qualify, let alone quantify, in isolation from other social, environmental and political changes. Thus, in effect, this chapter deepens the argument from the previous chapter – just having access to renewable energy sources does not necessarily mean more sustainability – given that energy practices encompass more than technical or material change. Only when environmental sustainability is seen as an integral part of the energy-modernity dialectic will it be possible to start identifying the opportunities for meaningful intervention.

In the first part of this chapter, I analyse how inequality at the village level translates into unequal access and use of electricity. Next, I analyse some transitions in the use of appliances, cooking equipment and agricultural technology. The second half of this chapter shows how these different technologies and appliances have different meanings for different generations, and create tensions between different cultures. The chapter concludes with an analysis of how these different elements lead to changing everyday energy-modernity in Southeast Asia, through the changing experience of distance and time, the shift from community to individual, and the increasing invisibility of energy and geography of cost and benefit.

The Relations between Access to Electricity, Assets and Income

Access and use of energy are often linked to economic development and wealth. Not only can increased use of energy lead to more productivity as often suggested (see Cabraal et al., 2005, for a review), but existing patterns of unequal distribution of wealth can also structure the uptake and use of different types of energy. Simply put, families with more income or assets are expected to spend more money on their energy services. The cases of Phakeo and Nam Ka show that there is a connection between wealth – in the form of assets such as livestock and land, and sometimes monetary income – and the adoption of electricity.

In Phakeo, the tariff system may be seen as a natural experiment which divides the population in three different groups. Table 5.1 shows these three groups and their relation to indicators of wealth and assets, such as land, livestock and rice sufficiency per household. The table shows that there is a plausible link between the assets of the households and the tariff. Households in higher tariff levels in the interview sample had, on average, more head of livestock and more land (both paddy and upland). Rice sufficiency seemed to play a role as well, as none of the households in the lowest tariff in the sample reported that they were rice sufficient. In contrast, monetary income in Phakeo seemed to bear no relation to the tariff level chosen, perhaps even a negative one. This could simply reflect the difficulty encountered when asking people for their income details (Deaton, 2005) or it could show the high variability of income from year to year of

Table 5.1   Tariff level, livestock and rice sufficiency in Phakeo

Tariff level	No. cows/hh	No. buffalos/hh	Paddy rice (ha/hh)	Upland rice (ha/hh)	Rice sufficient	Average income (LAK/hh/a)
1 (N = 4)	3.5	0.5	1.3	2.3	0%	8,100,000 (US$1,010)
2 (N = 5)	6.2	2.2	1.6	2.6	50%	15,900,000 (US$1,980)
3 (N = 9)	7.8	3.9	6.9	4.1	44%	6,700,000 (US$840)

Source: Author.

the people in rural Laos (Rigg, 2005). The qualitative interviews indicated that villagers' monetary incomes varied dramatically from year to year, for reasons such as marriage, sickness, diseases of livestock or crops, bad weather, and the need to take out or repay loans. This suggested that assessment of livelihood assets, such as land and livestock, is probably more important for determining the levels of electricity used.

In the case of Nam Ka, a similar point vis-à-vis the connection of monetary income and assets can be made through a similar natural experiment. Table 5.2 shows the average of assets owned by families that paid the connection fee for the rehabilitated micro-hydropower system in 2007, as well as of those who did not. This relation is somewhat less direct than the tariffs in Phakeo, as the situation for each household may have changed from that time (2007) to the time of data collection in 2011. Notwithstanding, the data show a plausible relation between the adoption of the hybrid hydro-solar system and the villagers' assets. While the amount of land between those with and without access to electricity at that time was not very different, the amount of upland rice in the category that did not get connected to the system was much higher. As discussed in the previous chapter, upland farming was generally undertaken by people who could not get paddy fields, either because they moved into the village late or because they could not afford to buy land. Most of the households that got an electricity connection through the rehabilitated system reported to be rice sufficient, while only approximately half of the 'non-adopters' said they were rice sufficient.

There are important differences in income between the households that adopted the hybrid hydro-solar system and those that did not. The key difference was that whereas only two households among the non-adopters said they gained income from livestock, all but two in the group of adopters claimed likewise. The situation in Nam Ka with regard to livestock was a little

Table 5.2   **Connection to Sunlabob, land, livestock and rice sufficiency in Nam Ka**

Connection to Sunlabob system?	No. cows/ hh	No. buffaloes/ hh	Paddy rice (ha/hh)	Upland rice (ha/hh)	Average income livestock (LAK/ hh/a)	Average total income (LAK/ hh/a)	Rice sufficient
Yes (N = 10)	7.1	3.8	0.8	0.1	3,800,000 (US$475)	4,491,000 (US$561)	89%
No (N = 6)	3.2	0.0	0.6	0.5	833,000 (US$104)	992,000 (US$124)	58%

Source: Author.

different from that of Phakeo in the sense that income from livestock varied from year to year. Therefore, monetary income on its own was probably not a very reliable indicator, supported by data from the qualitative interviews. A 91-year-old man in Nam Ka, for example, said that while the community built the system together, only the rich were eventually connected.

In sum, this section shows that there was a plausible link between energy transitions and wealth (albeit in different forms) in rural Southeast Asian villages, such as Phakeo and Nam Ka. In addition, livelihood assets are shown to be more important than monetary income. Apropos of the two case studies in Thailand, it was more difficult to make the link between energy and wealth since most people in these places could afford electricity and some electrical appliances. My observation in Mae Kampong, revealed that even though the villagers had small incomes and seemed to live in fairly poor conditions, they could still afford to buy new TV sets, either on credit or with support from relatives. Yet, the qualitative survey questions also showed that village-level statistics should not be taken at face value: many factors influenced peoples' decisions and practices regarding energy practices, factors often beyond their control that could change their livelihood situations quite dramatically. A family in Phakeo, for example, used to have two light bulbs, one for the house and one for the shop. After their daughter died, they had to close the shop and reduce the number of light bulbs and the expenses for lighting. The rest of this chapter will explore these kinds of changes further, mainly in relation to what people did with their energy once they gained access to it, starting with the relation between energy transitions and appliances.

Energy Transitions and Diffusion of Technologies

Access to electricity is often taken as an end-point or policy goal. However, for people who gained access to electricity, it is much more important to ascertain what they do with the energy once they access it. This section explores the adoption of different technical appliances and machines in the four field sites.

Comparing Appliance Use

The degree of uptake of appliances by the survey respondents in each of the four case studies is shown in the graph in Figure 5.1. Certain trends immediately stand out, such as the difference between Nam Ka and Phakeo with fewer appliances on the one hand, and between Mae Kampong and Bo Nok with a lot of appliances on the other. These differences were probably due to the longer time period of electrification, the higher levels of electricity available, and the higher levels of income and number of assets in the latter two case studies. Another feature that stands out is the high uptake of mobile phones across all four case studies, and television to a lesser extent, which gives some indication of the priorities associated with the acquisition of appliances. This also shows how some of the poorer rural areas in Laos have managed to catch up with – and even 'leapfrog'– some trends. This chapter explores these developments in more detail, gradually adding layers of complexity. The end of the chapter provides a comparative table (Table 5.5), which puts these findings into national perspective.

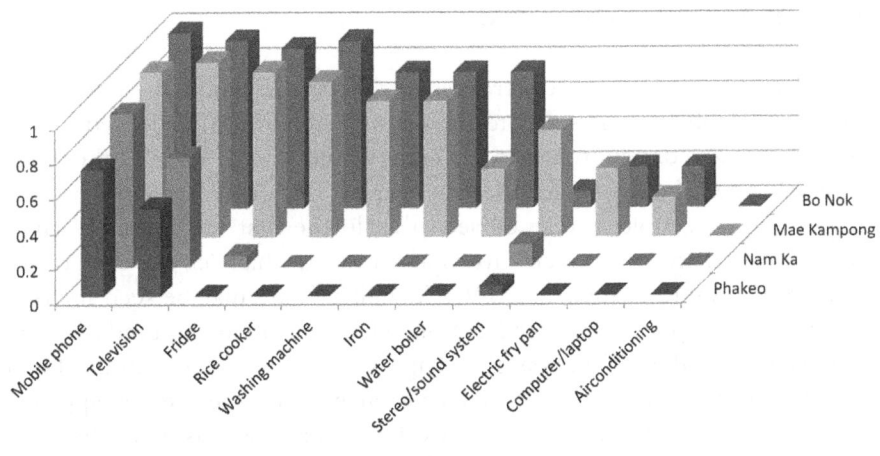

Figure 5.1   **Uptake of appliances in the four case studies**
Source: Author.

Figure 5.1 can be qualified by looking at the number of appliances per household in Table 5.3, which shows significant differences between each of the cases, also for television and mobile phones. While on average all households in the survey had more than one mobile phone, the number in Bo Nok was much higher than in the three other areas. A similar pattern may be seen for television sets. The interviewed households in Phakeo and Nam Ka had around one for every two households, whereas the households in Mae Kampong and Bo Nok had more than one and a half per household. Moreover, the larger household sizes in the Lao case studies further reduce the number of appliances per capita. Finally, the television sets that the people were currently using were bigger than in the past and (generally) consumed more electricity. While such characteristics are not captured in the graphs, they were confirmed though the qualitative data from the surveys and observations.

Table 5.3   Average number of mobile phones and televisions per surveyed household in each of the four cases

	Mobile phones	Televisions
Phakeo	1.3	0.5
Nam Ka	1.8	0.6
Mae Kampong	1.8	1.6
Bo Nok	2.8	1.7

Source: Author.

Uptake of Appliances over Time

There were some important differences in the ways in which these two main appliances – mobile phones and television sets – were being adopted in each of the four cases corresponding to the energy trajectory of each of the places. Figure 5.2 shows the figures for the first year of TV usage over time. For Mae Kampong, for example, it may be seen that in the first few years of having access to electricity from micro-hydropower, more than 65 per cent of the people interviewed got their first TVs. A similar figure may be seen for Nam Ka and Mae Kampong. Only for Bo Nok, where electricity has gradually been introduced since the 1970s, was the shape of the graph more gradual. The qualitative survey data suggest that television has become the key appliance for accessing entertainment, news and other information as a substitute for the radio. However, radio is still the preferred option among some groups, for example, the older people who used to work in the *miang* fields in Mae Kampong.

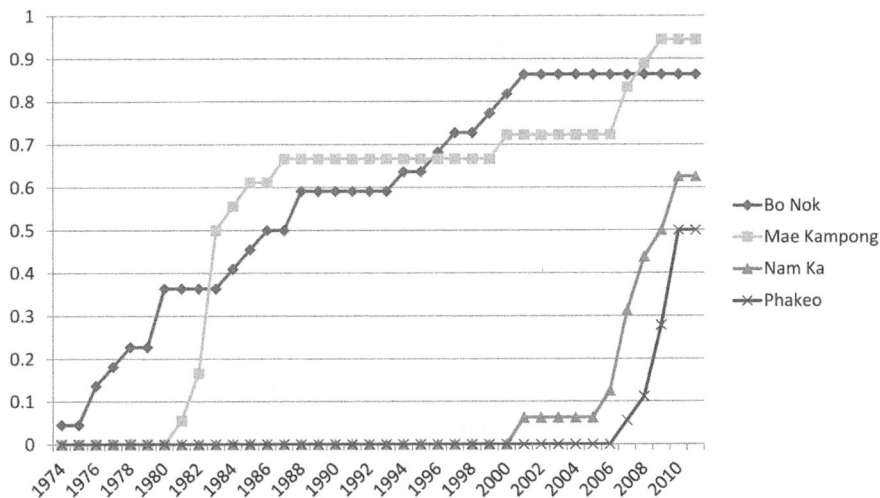

Figure 5.2   First year of television ownership of respondents in each of the four case studies

Source: Author.

Figure 5.3 shows a similar graph for the first year of uptake of mobile phones. As this device only became a mainstream appliance in the 1990s, the curves are much steeper than those for television. Moreover, they show considerably more similarities between the cases in Thailand and Laos, probably because of the lower costs and electricity use compared to TV. The rise in the number of mobile phones in Phakeo was particularly quick: in 2011, their uptake was almost similar to that in Bo Nok and Mae Kampong. The introduction of the solar mini-grid in Phakeo was the most likely reason for this rapid increase. Many people on the lowest tariff level in Phakeo could not recharge their mobile phones in their own houses: they had to go to their neighbours' or relatives' houses. However, the number of mobile phones per household and per capita is still lower in Phakeo than in the other three field sites (see Table 5.3).

Despite these seemingly uniform patterns, the reasons for acquiring appliances varied widely between the people surveyed. Some of them simply bought them because they wanted to have them or thought they needed them. Others purchased certain appliances because their neighbours had them. Some got appliances from their relatives. The mobile phone is a good example here: children like to give their parents mobiles so that they can keep in touch. Some people get their appliances for free with certain purchases such as fans with the purchase of a TV or a fridge. The timing of appliances uptake is often related to the particular stage of life people are currently in. Many people purchase their television sets, fridges or irons when they get married or build a new house.

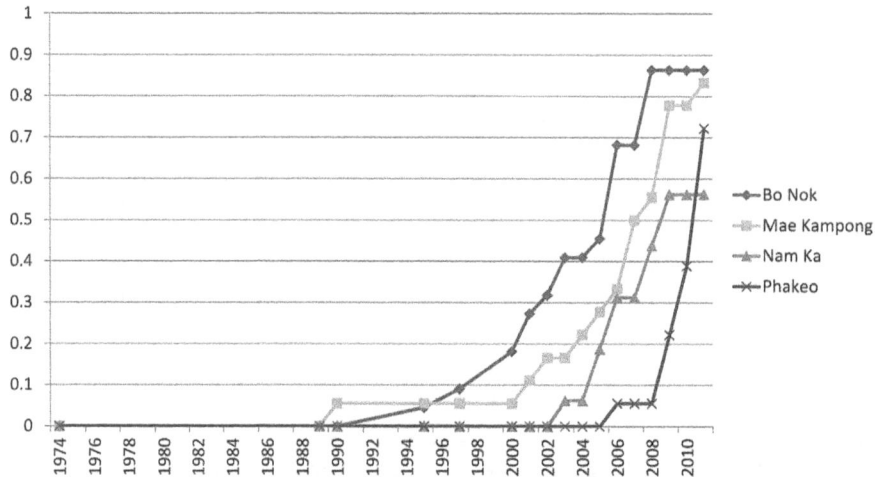

Figure 5.3   First year of mobile phone ownership of respondents in each of the four case studies

Source: Author.

Kettles, on the other hand, are often acquired when babies are born, and irons are needed for school uniforms. For all these reasons, energy appliances are connected to the different walks of life people pursue, and the associated energy practices reflect certain stages of their lives.

Transitions in Cooking Technologies

Examination of transitions in cooking practices provides insights into how these practices are related to transitions in electricity systems and how energy transitions work more generally. Figure 5.4 shows the cooking practices of the surveyed people in the four case studies. In Phakeo and Nam Ka, in 2011, the cooking done by people in the survey was still 100 per cent based on wood, which was collected from the nearby forests. Wood-based fires were used for space heating as well, because it can be around 0 °C in both Lao villages as well as in Mae Kampong during winter and in the early morning. Some people in Nam Ka and Phakeo said they would like to use gas or electricity for cooking, but cannot afford it at the moment. A 35-year-old man in Nam Ka believed that they will switch to electric cooking in the future, because firewood is becoming harder to collect. A man of the same age was under the impression that electric cooking would save him the time and energy he expended on collecting firewood and reduce the smoke coming from the wood fire.

Of all the cases, Mae Kampong was clearly in transition when it came to cooking practices. Traditionally, people used firewood to steam the *miang*

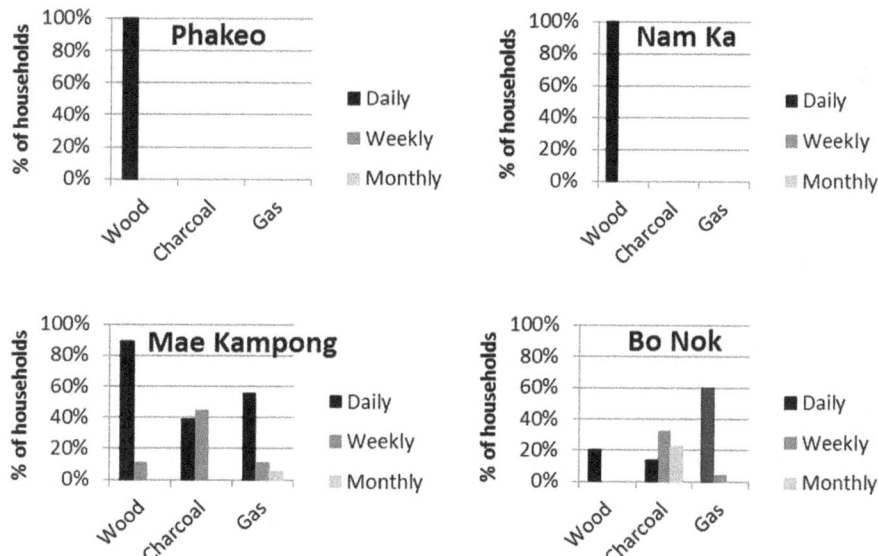

Figure 5.4   Energy sources and cooking practices of the respondents in the four case studies

Source: Author.

every evening (for sale), which explained the high number of villagers who used wood daily. The wood that was left over could be used as charcoal, which meant that most people did not have to buy it. Charcoal was used for grilling fish, meat, and other types of food. Gas had been used since around 1990: in 2011, approximately three-quarters of the respondents used gas (see Figure 5.5). However, some people, like the 43-year-old deputy village head, claimed that despite these transitions, his family's eating habits showed little change. They still predominantly ate sticky rice, but might use the rice cooker two or three times per week; for example, when tourists came for homestay. They only used gas to heat up food and for tourists' requirements. Electric fry pans were not very popular: the seven households that had one said they rarely used it.

In Bo Nok, finally, the survey respondents seem to have transitioned almost completely away from wood to charcoal, and mainly utilised gas for cooking. Moreover, in general they seemed to be cooking less and buying more food from stalls and restaurants. Furthermore, some Bo Nok villagers said that they had bought more efficient gas stoves in recent years, or were considering to buy one. A 28-year-old female resident of Bo Nok claimed that gas was cheaper to use than wood and charcoal, but the initial investment in an efficient gas stove was high. Another respondent said that she had just bought a more efficient stove for 3,000 THB (US$100), whereas her old one cost only 600 THB (US$20).

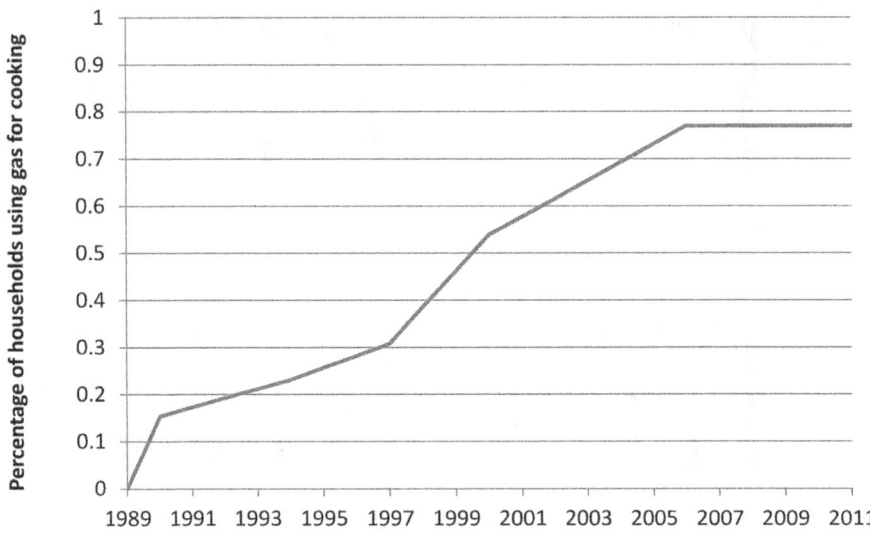

Figure 5.5   Year since first using gas for cooking in Mae Kampong (n=13)

Source: Author.

To refill an empty gas bottle costs around 330 THB (US$11), but the bottle itself costs around 1,580 THB (US$50).

My analysis of cooking practices across the four case studies provides some important insights into energy transitions on a household scale. First of all, these practices are changing slowly and the causes of change are manifold and context-dependent. Second, the transitions do not necessarily have to be from one technology to another. The examples of cooking practices show that different ways of cooking exist parallel to each other, such as in the case of Mae Kampong. Third, the transitions in cooking practices have gender-specific impacts. As cooking in Laos and Thailand is predominantly done by women, they are most likely to change their practices as a result of the use of different cooking technologies, rather than those of men.

Transitions in Agricultural Machinery and Agricultural Practices

Another example of non-electricity related transition is the use of agricultural machinery, such as tractors, which has had important implications for the agricultural practices of the peoples of Southeast Asia. One of the key changes in rice cultivation has been the introduction of the two-wheel or single-axle tractor, which has been dominant in rural Thailand for decades (Baker and Phongpaichit, 2014) and is becoming increasingly prevalent in Laos. Tractors are mainly used to plough the paddy fields, replacing buffalos as the key source

146

of power. Other activities, such as planting rice, are often still done by hand, although planting machines are becoming increasingly common for larger plots, increasing also outside Central Thailand. These tractors can also be used for other purposes; for example, to transport goods and people, collect firewood, even to generate electricity with a separate generator. Generators are particularly important in places with limited electric power, such as Phakeo, for use with sawing machines, air compressors and welding machines.

In Nam Ka and Phakeo, where cultivating rice is a key livelihood activity, the use of the tractor is a fairly recent phenomenon. In a period of less than 10 years, tractors have become a common technology among the surveyed peoples, with an uptake of around 30 per cent in Phakeo and around 40 per cent of all respondents in Nam Ka. While the adoption rate in Phakeo started later, it caught up rapidly in this place. The total number of people actually using tractors is much higher than the ownership rate, as it is very common to borrow or rent tractors within or even from outside the village. Renting a tractor costs between 300,000 and 600,000 LAK (US\$37.5–75) per year, excluding costs for diesel of around 700,000 LAK (US\$87.5). In comparison, one family in Phakeo that bought a new tractor in 2009 for 11.4M LAK (US\$1,425), can make part of this substantial investment back by renting their equipment out to other people. However, the demand for renting tractors in Phakeo has decreased, because there are many in the village now.

In Mae Kampong, no mechanisation of agricultural activities has occurred, because the production and picking of *miang* and coffee is done entirely by hand. Tractors are not allowed in the forests surrounding the village. However, some indirect benefits flow from the use of vehicles such as transporting produce from the village to the market. In the past, the Mae Kampong villagers used cows, each of which could carry 200–300 bunches of *miang*. These days, according to an older Nae Kampong couple, people drive their own trucks, or the bunches get picked up by middlemen.

The villagers' livelihoods and agricultural practices are highly diverse in Bo Nok, making it more difficult to make general statements about the mechanisation of agriculture. In general, the agricultural activities are large-scale and dependent on the use of trucks and other machinery. Shrimp farms, for example, are highly energy intensive, but also need large amounts of labour at the time of the capture and sorting of the prawns. In the past, all the shrimp farms used diesel engines to drive their paddlewheel aerators, but, nowadays, some are connected to the electricity grid. Coconut, palm oil, aloe vera and various fruit plantations are labour-extensive during the time of growth, but need to be harvested by hand. Similar to the case of Mae Kampong, the transportation of all of these crops has benefited from the use of (pick-up) trucks.

Technological Change and Implications for Energy Use and Environmental Sustainability

The general trend towards the use of more appliances and machines has led to more energy use, which generally has negative implications for environmental sustainability. Yet, this relation is not straightforward because it depends on the nature of the appliances, the ways in which they are used, and their sources of power. In other words, it depends on the change in energy practices. In Mae Kampong, for example, people used electricity generated by their own micro-hydropower exclusively until 2000, a practice that did not negatively impact on the environment. Thus, one could say that before 2000, it did not matter how many appliances people used – subject to the availability of enough water – as it would not lead to more fossil fuel use or pollution. However, the increasing uptake of different appliances has laid the basis for increased demand for electricity and eventual connection to the grid, for which the source of electricity is far away and often based on fossil fuels. Lighting is another important example which shows the complexities of technology use and sustainability. As in many other countries, the incandescent lamp has slowly been phased out in Thailand and Laos and replaced by more efficient compact fluorescent (CFL) bulbs and tubes. In some areas such as Phakeo and Nam Ka, for example, people never used incandescent lights, as they have only recently been able to access electricity. In Bo Nok and Mae Kampong, most of those interviewed had already replaced their incandescent lights and only very few indicated that they are still using some. Thus, one could say that the phasing out of incandescent lights has led to less electricity use. At the same time, however, the number of lights has increased in peoples' houses over time. And, while it is difficult to quantify this change, many people said that they have more lights now than they had in the past. In some cases people went from using one or two incandescent lights two decades ago to using more than a dozen CFL tubes now, resulting in an increase in actual electricity consumption. Moreover, in many places in Laos, the CFL lights are low in quality and need frequent replacement, leading to increased costs and reduced environmental sustainability.

In addition, most people use their lights longer than they did in the past. In Nam Ka, for example, where electricity used to be very expensive under the public-private Sunlabob system, people used their electric light sparingly to save costs and considered this 'normal'. Nowadays, people use the same number of lights for a much longer period of time. These changes in the construction of 'normality' in terms of levels of illumination are more important than technological change in itself to understand sustainability transitions, which is a key point of the energy practices literature. This change in routines can be illustrated with a practical example. According to a 56-year-old man and a 54-year-old woman in Mae Kampong, electricity and electric light are very important for the processing of *miang*, which usually happens at night. Before they had electricity, people had to use candles, which provided much less illumination.

Whereas the contribution of lighting in rural areas is minor in the bigger picture of increased energy demand, these processes and trends provide important insights, especially when extrapolated to other technologies such as air-conditioning, because the latter uses a lot of electricity compared to other appliances. Widespread use of air-conditioning will lead to increased demand for power plants and, by extension, to a less sustainable energy system. In the four case studies surveyed for this research, none of the respondents indicated that he/she had air conditioning at home. This is not surprising given the generally low uptake of appliances in Nam Ka and Phakeo and the relatively cool climate in Mae Kampong. In the case of Bo Nok, however, a few people expressed interest in buying air-conditioners: some households already had one, including that of one of the prominent figures in the Bo Nok anti-power plant movement. The latter is a good example of how the discourse of local conservation does not always match individual behavioural patterns, a phenomenon common in developed countries, but which can increasingly be seen in middle-income countries. Moreover, short interviews at two electric goods shops in the provincial capital revealed that the demand for air conditioning has grown rapidly since they first started selling them around 2006. Whereas nowadays these shops mostly sell them to people in the urban areas, they said that they also sell some to those living in the countryside. With the relative price of energy decreasing and the fact that – according to some of the respondents – the climate has become hotter in recent years, this trend is likely to continue. With the introduction of more air-conditioned spaces, such as shopping malls and restaurants, perceptions of a 'normal' or modern indoor climate in Southeast Asia will also contribute to this trend.

To conclude, clearly there are relations between increased use of appliances and environmental sustainability, but they are far from linear or straightforward. Moreover, analyses of these complex relations often fail to capture the changes appertaining to the meaning and competences of people, rendering them less suitable for understanding how energy affects social change and modernity. The energy practices framework of this chapter suggests that the construction of normality in terms of appliance use and machinery provides a better explanatory framework for both the long-term and everyday implications of energy transitions, as the above examples show. In the next section, focus is upon how practices differ between different generations, age groups and ethnic groups as a result of the energy trajectories.

Energy, Generations, Culture and Changing Worldviews

Television and Experience of the World Outside

Television is one of the most important ways in which people gain knowledge about the world outside of their villages, especially in remote areas where people have limited opportunities to travel. In all four case studies, television is a very important object in the household, evident in the fact that it is usually placed in the middle of the living room. As suggested earlier, it seems that television has taken the place of radio as the most important source of news and entertainment in most households. Both in Thailand and Laos, people mainly watch Thai television, for example news, soap operas, sports and cartoons for children. Music videos (VCDs) are also very popular, either in Thai or Lao language. In villages with ethnic minorities, such as Nam Ka and Phakeo, people watch videos in their own languages. The use of a satellite dish to receive television is necessary in some areas in Laos, but has also become popular in other areas, such as Bo Nok, as villagers seek to access more channels.

People expressed different opinions about whether television has actually changed their lives or not. In Nam Ka, there was a strong division between those who said they learn a lot from TV and those who simply watched for entertainment claiming that they do not learn anything. For example, one 35-year-old man in Nam Ka said that TV has improved their lives in many ways: they have learned to dress from soap operas and can learn the Thai language. He was also interested to learn other things from foreigners through television. Even people who did not have a TV thought it could change their lives. A 63-year-old man from Nam Ka said that he was dreaming about having his own TV, and claimed that he would watch day and night when he eventually gets one. He would like to see other countries, big cities, and to learn how people behave. Moreover, he thought that the people in movies are nice and try to understand each other, which is something people in the village could learn from. Other people, for example, a 63-year-old female, were less optimistic about being able to learn from television. According to her, the lives of people on television are very different from theirs and people like her cannot follow what they are doing. Finally, some people said that television is simply for entertainment and does not actually influence their lives in any way.

Children were very attracted to watching television in all four places, an impression that came from interviews as well as from observations. Many could be seen watching television, often for a substantial number of hours per day. Again, opinions regarding the influence of television on children were divided. One of the teachers in Phakeo, for example, observed that watching TV has caused big changes in the way children learn. Since they have been watching TV, children have become cleverer and faster, have more knowledge, know the news and are less shy, in general. He observed that sometimes children under five years old had already started to remember things they saw on TV. A 33-year-old man in Phakeo thought that TV could help children to understand aspects of the economy, development and politics in Laos. Moreover, he did not think that watching TV interfered too much with their homework as was sometimes suggested by others. Some people, for example a 46-year-old female in Nam Ka, emphasise the fact that television can provide role models for children. She said that on TV, all of the people are highly educated and have an easy life, so she told her children that if they have a good education, they can live an easy life too. A 28-year man from the same village put it even more directly when he said that if children are not good at studying, they will have to spend their lives farming. If they prove proficient, however, they can find a job in town. However, not everyone was so positive. One of the former village heads in Mae Kampong thought that children who watch TV change their eating habits, orienting them towards snacks and junk food. He was also afraid that soap operas might teach children about drinking, smoking and other negative things.

The first key point that emerged from these comments was that, for some, television was seen as an important vehicle to learn about the world outside and a way to change their lives. These comments were more common in Nam Ka and Phakeo, where TV is still somewhat of a novelty. For villagers from Nam Ka and Phakeo, what can be seen on TV has an important role to play in shaping the idea of modernity to which many people seem to aspire. Importantly, this notion of modernity is at least in part based on Thai urban life, as much of the television that people watch comes from metropolitan Thailand. In the two Thai case studies, more people emphasised the potentially negative influence of television. This response may have been related not only to the longer exposure to television in Thailand, but also to the prevalence of more critical discourses of modernity in this country in general. The different influences of TV and other electrical appliances become more pronounced when differentiating between different groups of people, which I will attempt in the next few sections.

Energy Practices in Different Generations

Due to the rapid changes in energy-modernity in many parts of Southeast Asia, energy transitions have different generational impacts. One of the most important differences is between those who grew up with electricity and electric appliances and those who did not. Moreover, younger people often have more experience of travelling or living outside of their villages. The story of a 19-year-old student from Mae Kampong illustrates some of these differences. At the time of interview, this student was in her second year of Library Studies at Rajabhat University in Chiang Mai. Even though she had not been away from her village for long, she was already more familiar with 'urban lifestyle' than her parents. She liked to take hot showers, whereas her mother takes them cold. She was also more familiar with cooking with gas, rather than charcoal or wood. And while she has helped her parents a few times to pick *miang*, she is not very good at it. She said she liked the village for its climate, but she was not sure whether she would be able to find a job there. One of her few options was to do something with tourism. When asked what she would wish for the future, she said she wished that all of the phone companies will have signal in the village as well as wireless internet.

This example shows how the different paths of life have important implications for the ways in which energy is used and perceived. There are many other examples which point to the fact that while electricity has become a given for many of the younger people, their parents often perceive it as a luxury as many did not grow up with it. The example of the hot shower, which is required for houses that offer home stay in Mae Kampong, is a good example here. Many interviewees said that they only provide it for their guests, not for their own use. While this is partly a way to save costs, it is also partly a lifestyle that showcases a difference between generations.

The transition in cooking practices, from wood and charcoal to gas and electricity, is another aspect of the story of the student in Mae Kampong. While many of the villagers in Mae Kampong and Bo Nok had gas stoves, they often still preferred to use their traditional ways of cooking. A 65-year-old woman from Bo Nok, for example, said that she cannot get used to gas, even though she has a gas stove at home. She said that the flame is always either too high or too low and she has burned herself many times. Other interviews confirmed that gas is used as supplementary by many older people; for example, for heating up food rather than the primary way of cooking, as was the case of the student in Mae Kampong and other adolescents studying in the city. There were some exceptions such as a 73-year-old man in Bo Nok who said that he used to be scared to use gas, but not anymore. He now uses it more and actually thinks it is easier and quicker than firewood or charcoal.

The use of mobile phones is another important case in point of how practices related to energy differ between generations. While most people in Thailand and Laos, both young and old, are by now familiar with the use of mobile phones, many of the older generation only use them because their children gave them one to stay in touch. A 55-year-old woman in Bo Nok, for example, said that she only answers the phone: she never calls anyone herself. For younger people, the mobile phone is not only used to talk to friends, but also to take photos and listen to music. For example, one 22-year-old fisherman in Bo Nok uses his mobile phone to listen to the radio, even though they also have a regular radio at home.

A final example of how different generations can have different energy practices is mobility. In general, practices of mobility vary considerably depending on the stage of life of the people concerned. Usually, those in their early working lives travel a lot for work, family and other activities. Older people I interviewed usually say that they travel less nowadays, because they do not need as much and it is more tiring for them than in the past.

Energy, Youth and Modernity

Energy transitions can create or exacerbate tensions between everyday life and promises of modernity, especially for younger generations. In Laos, as in many other low and lower-middle income countries, there are high numbers of children and youth due to growing prosperity, the reduction of diseases and child mortality. Thailand is in a different stage of demographic transition given its increasingly aging population and rapidly shrinking rural population. In both countries (Laos and Thailand), however, the younger generation have different hopes and expectations from their predecessors. Many of these changes may be related to changing energy practices, such as improved communication, access to information, and transport opportunities.

In Thailand, many people aspire to find jobs outside of agriculture, in particular the younger generation in peri-urban areas such as Bo Nok, as well as in well-connected and prosperous villages such as Mae Kampong. Many of these jobs would not be possible without access to energy and transport. Moreover, they are linked to a particular discourse of modernity, which is almost invariably associated with an urban rather than a rural lifestyle. The traditional agricultural livelihood, which involves long days of working in the fields, is seen as hard and difficult by the younger generation. Moreover, manual labour and working under the sun has a lower status than working inside, in factories for example. The increased income, mechanisation and growing importance of education means that children now have more choices when they grow up. A 60-year-old man in Mae Kampong said that children in the village used to start picking *miang* when they were eight or nine years old, but now they are all at school at that age. After they finish school, many younger people have the opportunity to find jobs outside of agriculture, sometimes even outside of the village. Examples from the surveyed people in Bo Nok and Mae Kampong include people that have become shopkeepers, drivers, construction workers, traders, waiters and various tourism-related positions (Figure 5.6).

In Phakeo and Nam Ka, however, there are generally fewer opportunities for people to find jobs outside of agriculture. The villages are more remote and there are not many places that offer alternative employment. In addition, peoples' economic situations and education levels make it more difficult both to travel to the cities and to find jobs. In Phakeo, some men do construction work

Figure 5.6 Adolescent in Mae Kampong using his laptop
Source: Author.

on projects such as the nearby Nam Ngum 5 hydropower dam. Other than that, most of the additional income-generating activities in Phakeo and Nam Ka are supplementary to the traditional agricultural work. Examples from the survey include small shops, handicraft, traders and middle men, scrap metal collection, and transport services.

Conversations with groups of young people (between 15 and 30 years of age) in Nam Ka and Phakeo also revealed some of their aspirations and their ways to negotiate tradition and modernity. On the one hand, they saw their lives as different from those of their parents, because today they can drive their motor bikes to nearby towns, charge their mobile phones in the village, wear 'Western' or low-land Lao clothing, and are happy to watch as much television as they can. On the other hand, they still have to help their families with rice cultivation, have to gain approval from family and friends to marry the 'right' person, and know that most likely they will spend their lives living in the village. The fact that both the Phakeo (Khmu) and Nam Ka (Hmong) peoples are distinctly different from their dominant ethnic lowland Lao counterparts contributes to these tensions.

In all four case studies, the distance between the village and the rest of the world has become smaller, both physically through the improvement of roads and transport opportunities, and mentally through television, mobile phones and other appliances. In Laos, however, the contrasts between the everyday routine on the farm and the images on television are probably even stronger. The combination of this contrast and fewer opportunities to find jobs outside of farming can be seen as a dilemma or tension of modernity related to energy trajectories. These tensions may be detected in the differences in the clothing and hair styles between the young and older villagers, and in their attitude towards their ethnic cultures in villages like Phakeo and Nam Ka.

Tensions between Ethnic and Mainstream Culture

Transitions of energy systems may add to tensions between traditional and perceived mainstream 'modern' cultures, in particular regarding issues of ethnicity and religion. Nam Ka is one example where the contrast between the traditional ethnic culture and mainstream lowland Lao culture is clearest. When asked their opinions, people in Nam Ka seemed to think that their lives had changed profoundly as a result of television and increased opportunities to travel. According to a 35-year-old male villager, people in Nam Ka have 'dropped everything' cultural, except their religion. He said that the past is not important and that everything changes, year by year, century by century. A middle-aged woman explained that she bought a TV because they want to learn about being modern; for example, through observing the clothing in soap operas and things from the news. She continued, saying that she does not want to follow the traditional practices any longer since she has seen the new things on television.

Other observations and follow-up questions revealed a more complex picture of how energy-modernity and the traditional Hmong culture are related. When looking at clothing, for example, few people continued to wear traditional Hmong clothing, except for some elderly people. The younger people preferred either Western-style clothes or low-land Lao clothes, in particular when they go to nearby villages or towns. They said that they do not want to be seen in their traditional clothes, because this would make them look 'backward'. However, they all kept their traditional outfits in the wardrobe to wear at traditional festivals such as the Hmong New Year. These examples show how the people of Nam Ka are trying to juggle external influences that come along with energy-modernity discourses, with their traditional culture.

Another interesting aspect of energy and culture is how energy-modernity and new technologies can reinforce particular cultural values, while at the same time introducing new ones. A key example is the popularity of Hmong-language VCD movies in Nam Ka. These videos, which are often produced in Thailand – where there is a sizeable population of Hmong people – are sold in the nearby district centre or in the provincial capital. There are hundreds of videos, ranging from music to movies in all different genres. My observations in Nam Ka indicated that some of the drama/comedy-type movies bearing a strong moral message (delivered in a light-hearted way) were amongst the most popular. Some of the messages in these movies are about traditional values, such as being honest, respecting others, and about traditional Hmong practices. However, there is evidence of elements of change in some videos, which show, for example, that one wife provides enough trouble, a concept at odds with the Hmong tradition of polygamy. A 22-year-old man said that they get other values from these videos too, such as the need to try to understand each other and deal with conflict situations.

Energy Transitions and Everyday Modernity

In the last section of this chapter, the trends identified in the first part are related to the ways in which energy transitions influence everyday experiences of modernity.

Experiences of Distance and Time

Energy transitions play an important role in the changing experience of distance and time, which Giddens (1990) recognises as an important aspect of modernity. There are a number of different elements that contribute to this changing experience of time and distance, which can be grouped together under mobility and communication practices, both of which are often literally powered by the ongoing energy transitions analysed in this chapter. These changes are discussed using quantitative data first, followed by a discussion based on qualitative (interview) data.

In the remote rural areas of Southeast Asia, people used to walk and cycle everywhere for their work and other needs, sometimes until fairly recently. This has changed drastically with the introduction of tractors, motor bikes, cars and other vehicles over the last few decades. Figure 5.7 provides an indication of how these changes developed for each of the four cases by looking at the adoption and penetration of motor bikes over time. The pattern in this figure is similar to other graphs in this chapter: it shows a gradual increase of motor bike ownership amongst the respondents in Bo Nok since 1970. In Mae Kampong, the first respondent acquired his motor bike around 1990 after which there was a rapid increase to 100 per cent of the sample by 2008. Nam Ka and Phakeo showed similar trends, with the first motor bikes bought by respondents just before 2000, catching up quickly with over half of respondents owning one or more motor bikes by 2011. The number of cars shows a similar pattern, as may be seen in Figure 5.8. The total number of cars in each place was much lower compared to motor bikes though.

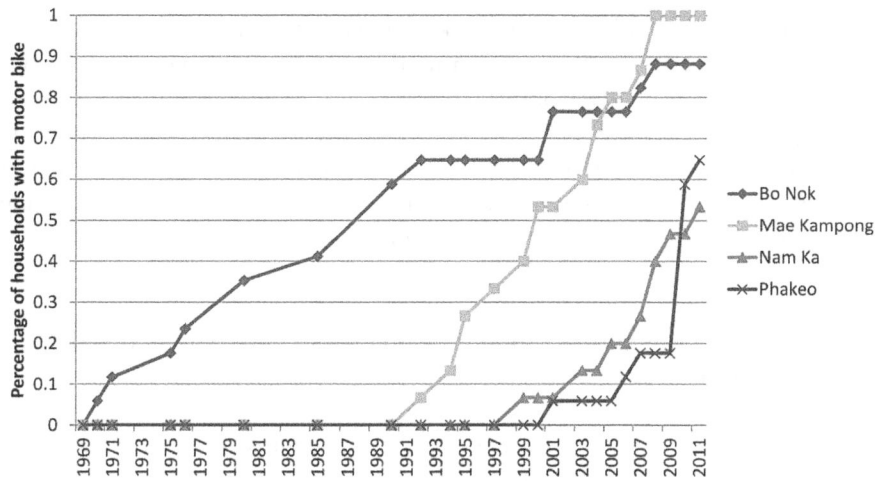

Figure 5.7 **Year of first motor bike ownership amongst surveyed households in each of the four case studies (n=65)**

Source: Author.

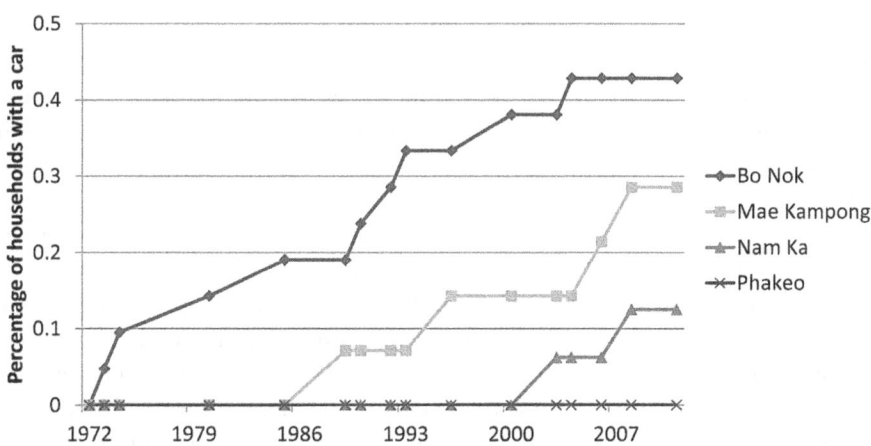

Figure 5.8 **Year of first car ownership among surveyed households in each of the four case studies (n=69)**

Source: Author.

In addition, the number of vehicles per household has also increased. Table 5.4 quantifies this change by showing the average number of vehicles per surveyed household in 2011. This shows that while the people of Mae Kampong and Bo Nok have at least one motor bike per household, on average, those in Phakeo and Nam Ka have just about one full vehicle (tractor or motor bike) per household. A 55-year-old lady in Bo Nok said that in the past, having one motor bike was already a lot, but that people nowadays have multiple vehicles so it has become much easier to get around. However, simply possessing vehicles does not change one's transport practices. For example, a couple in Mae Kampong (54 and 56 years old) had two motor bikes for their children, but they could not drive them themselves. This example stresses again the importance of changing practices, rather than simply counting the number of vehicles.

Table 5.4   Average number of motor bikes and cars per surveyed household

	Tractors	Motor bikes	Cars
Phakeo (n = 18)	0.3	0.7	0.0
Nam Ka (n = 16)	0.4	0.6	0.1
Mae Kampong (n = 18)	0.0	1.3	0.6
Bo Nok (n = 22)	0.0	2.2	0.7

Source: Author.

Some of the qualitative survey data provide more insight into how individuals experience these changes in transport options. In general, most people stated that they travelled more and further than they did in the past. A 78-year-old man in Phakeo, for example, said that before resettlement, his village had no road, so it would take almost a full day to walk to his upland rice field. Now, it only takes one hour to go to the district capital of Phoukhoun. A 47-year-old woman from the same village recalled that she would only go to the market perhaps 10 times per year to buy things like salt and MSG, because it would take nearly one full day on foot. Now, if they have money, they can go to the market every day by motor bike or tractor. In Bo Nok, some people stressed how services, for example health care, seemed to be much closer now than before. A 48-year-old woman recalled that in the past they had to go by cart and then train to visit the doctor, but now they have their own transport and can drive there whenever she chooses. Another interviewee, a 35-year-old woman, said that she goes to the hospital in the provincial capital these days, whereas in the past she used to go to the health clinic in the district capital of Kui Buri. Personal transport has made it easier for her to travel further and to receive better treatment. Other people said that it has become easier to go to the district or provincial capital for administrative issues such as going to the bank, attending meetings and training, and obtaining official documents.

Some of the older forms of transport such as bicycles, while still used by some, are generally seen as backward. For example, a 55-year-old woman in

Bo Nok said that she still had some bicycles, but did not use them anymore because they are not modern (*than samai*). She added that she actually wanted to sell them, but no one wanted to buy them. Another (54-year-old) woman from Bo Nok claimed that life was more difficult in the past because they had to ride their bicycles to work on the farm. Now, she said, people use their motor bikes everywhere to the point that she wondered if people can still walk!

Another way in which the experience of time and distance has changed is that motorised transport has become more individualised, rather than collective. In villages like Nam Ka and Phakeo, the first people to obtain motor bikes or trucks used them to take people to the district capital or to cart heavy goods (and charge people for the service). An example of this was a 28-year-old Nam Ka man, who bought a truck with support from his cousins in the US in 2003. He obtained a license from the government to start a transport service, to bring people to Phonxay and Phonsavan five times a week. In recent years, however, he has had difficulty finding enough passengers, because many people now have their own tractors and motor bikes or use other means of transport. A 37-year-old female in Mae Kampong said that before, people would share a truck or pick-up, but, nowadays, people just go by themselves. In other words, transport has transitioned from walking to shared and intermittent motorised transport to individual transport whenever needed. In some cases, as in Nam Ka for instance, this happened in the space of only one or two decades.

It is important to note here that while transport has become easier in many ways, it is by no means ubiquitous and accessible to everyone, reflecting both existing and new patterns of inequality. Many people in Phakeo and Nam Ka still go on foot to collect NTFPs or firewood, or walk to their rice fields, which can take up to half a day. Some Nam Ka villagers still have to walk for a whole day to get to their livestock, and there are no roads leading there. Also, in Mae Kampong, some people have to walk for around 1.5–2 hours to reach to their *miang* or coffee fields on very steep slopes when there are no alternatives available. Another constraint on the use of motor bikes and cars is the price of diesel, which has risen sharply over the last decade.

The experience of time and distance has also changed through the use of mobile phones, which sometimes replace the need for transport. For example, a 50-year-old female from Nam Ka said that her mobile phone has brought important changes to her life because she no longer needs to walk everywhere if she wants to pass on a message. A 35-year-old female from Bo Nok stressed the petrol she saves by using her mobile phone: she said that she no longer had to drive to see other people. The use of mobile phones has also changed the way people in the four field sites trade and do business. A 48-year-old man in Phakeo said that they can call to Kasi district – which is approximately 40–50 kilometres away – to order rice if they run short. In Bo Nok, a 55-year-old female described how mobile phones have changed the way she does her business of selling pineapples and mangos. In the past, she had to travel by bus to make deals, but now she can just phone people. She also uses her mobile phone to talk to her workers.

A third way in which the experience of distance and time has been altered is through the use of certain electric appliances such as fridges, freezers and rice cookers, which enable people to manipulate their time and daily routines. Warde et al. (1998) term the use of such appliances 'hypermodern' practices: people can save food to eat later or cook rice at any time without having to constantly pay attention. A 55-year-old woman in Bo Nok said that one of the key advantages of the rice cooker is that she can just plug it in and come back later, whereas before she had to sit, wait and keep tending the fire when she cooked rice. In some cases, hypermodern practices can reduce the amount of time spent on certain activities. For example, another 55-year-old lady from Bo Nok claimed that electricity has changed her life completely: she finds that things can be done much more quickly than before. She now gets up at 6 a.m., instead of between 4 and 5 a.m.

Whereas some of the implications for environmental sustainability of changing experience of time and distance are relatively straightforward, others are more complex. The increased use of motorised vehicles over longer distances means more fossil fuel usage and negative environmental impacts.[1] The implications of the use of mobile phones are complex: it could reduce physical movements and does not use a lot of electricity compared to the energy needed to drive a vehicle. However, the anecdotal data in this research do not suggest that the use of mobile phones leads to a decrease in distances travelled. On the contrary, research by Yuan et al. (2012) undertaken in China shows that mobile phone use is positively related to the distances people travel. Finally, the use of hypermodern appliances (such as rice cookers) generally lead to higher electricity consumption, because of their ability to be available on-demand.

Shift from Community to the Individual

Energy transitions also influence the balance between communities and the individual within each of the case studies. This chapter has already discussed some examples of how this balance is shifting away from the community and communal activities towards more emphasis on the individual, such as the use of individual modes of transport. This is an important shift in the experience of everyday modernity of people in Southeast Asia.

Besides transport, television is another important example of how the balance between the community and the individual changes. In all four areas,

---

1 The transition towards biofuels in Thailand is still marginal and controversial in itself.

the first arrival of electricity brought great excitement, and one of the first appliances of choice would be television. Among the surveyed respondents, half of all people (51 per cent) indicated that they bought a television in their first year of access to electricity. Moreover, many told stories of how the first houses to have electricity and television in the village used to be very popular during the evenings, becoming places of social gathering. A 60-year-old woman in Mae Kampong recalled that a lot of people came to watch television at her house, because it was the first in their hamlet (Pang Nai 2). At the time, her family bought the TV and the generator together for the substantial sum of around 30,000 THB (approximately US$1,000). In Mae Kampong, which is quite spread out, there was at least one household with a generator and TV in each of the six hamlets before the micro-hydropower system was established. Many of these places with generators would collect money from each of the visitors to share the costs of petrol.

Nowadays, many households in rural Laos and Thailand have TV-sets of their own as a result of increased access to electricity, higher disposable incomes (a fact that most people mentioned in interviews), and a decrease in the price of appliances. People like to watch television in their own homes without the need to share it with others. Thus, while early television adopters provided a platform for more focus on the community, watching TV has lost much of that function nowadays. In some ways, satellite TV, which is not available to every household, still attracts some crowds every now and then, albeit to a much lesser extent. One 28-year-old man in Nam Ka said that he watches satellite TV at other peoples' houses, because it is too expensive for him at the moment.

Some of the interviewees actively commented on the changing balance from community towards more individualisation. An older couple (54 and 56 years respectively) in Mae Kampong recalled that in the past their children would play and hang out together near the house, whereas nowadays they go everywhere by motor bike and car. A 62-year-old woman from Bo Nok observed that people used to work as exchange labourers on farms, but they ceased this practice some 10 years ago. In the past, therefore, farming was generally an activity that involved the whole community. Families would take turns helping each other, as still may be seen in the two Lao cases. However, the introduction of tractors has influenced this system too. Families can now rent tractors in the village to replace some of the manual and communal labour. A 40-year-old man in Bo Nok complained that life in the area is now like life in Bangkok; people might live next to each other, but they do not really know each other or what they do. Moreover, he added, people are quite self-centred now. There are fewer festivals, and people no longer really know what the village head is doing.

While some of these statements could be interpreted as nostalgia, they do point to some important shifts in the experience of the story of modernity and diffusion of energy, appliances and vehicles in each of the four cases.

Moreover, in general, the shift from community to individualisation has led to higher energy consumption, as the same resources are now often used by fewer people.

Invisibility of Energy and the Geography of Cost and Benefit

The energy trajectories in the four case studies show that electricity systems and other infrastructure have become increasingly taken-for-granted over time. This change in the status of electricity may be seen as one of the paradoxes of energy-modernity. While the demand for energy goes up, and the economic, social and environmental costs of producing energy increase, energy and electricity in peoples' everyday lives increasingly move to the background. This is particularly so in places that have had access to electricity for longer periods of time, such as Bo Nok and Mae Kampong. A 54-year-old man in Bo Nok commented that only when the electricity is cut off, he notices how much he has become used to it. In Mae Kampong, one of the key reasons why people to switched to grid electricity was that they could no longer tolerate the instability of micro-hydroelectricity. Energy infrastructure in these places is now only visible when there is a problem (Star, 1999); in other words, the availability of 24-hour stable electricity has led to a new level of normality (Shove, 1997). The same process is happening in Nam Ka, where people have recently been connected to the EdL grid and will soon get used to having 24-hour stable access to electricity. In Phakeo, they are constantly reminded of the limitations of electricity production when the system is switched on and off every day.

One of the implications of the invisibility of electricity infrastructure is that it widens the geography of cost and benefit (Hirsch, 2007), which is another key feature of contemporary energy-modernity. In the past, energy production and consumption were closely related and confined to a fairly limited geographical area; for example, the collection and burning of firewood. Decentralised generation, such as micro-hydropower in Mae Kampong and the solar village grid in Phakeo, shares some of the same features, as the electricity is produced and consumed within relatively small geographical areas. When places are connected to the main grid, however, they are no longer directly connected to the process of electricity production, which changes the geography of energy transitions (Bridge et al., 2013).

One of the consequences of the increasingly stretched geography of costs and benefits is that the influence on environmental sustainability becomes more dispersed and complex. An example from Mae Kampong before it was connected to the national grid illustrates this point. When there was a shortage of electricity in 1994, the village cooperative came together and traced the problem to the degradation of the watershed. They decided that in future, people would need permission to cut trees in that area. In other words, the influence of environmental degradation on electricity generation was immediately felt and resolved by people in the village. Since Mae Kampong has become grid connected, electricity is less a part of the villagers' livelihoods and environment – as is the case in most other places in Southeast Asia – and there are fewer incentives to balance electricity production and consumption, as the effects are being felt elsewhere.

The Changing Experiences of Energy-Modernity in Southeast Asia

The main conclusion of this chapter is that transitions in household energy end-use technologies co-evolve with changing livelihoods, worldviews, culture and practices, often in non-linear ways and with different impacts for different generations, levels of wealth and ethnic groups. Importantly, this chapter also shows that energy and modernity are interrelated and that the effects of transitions go beyond narrowly defined sustainability concerns, and include influences on equity, world views, culture and social relations. These contingent effects should be taken into account when analysing the transitions in end-use energy technologies. Moreover, the effects of the transitions discussed in this chapter on environmental sustainability have been demonstrated to be difficult to qualify, let alone quantify. Nevertheless, Table 5.5 consolidates the findings from this chapter. Focus is upon the key end-use energy technologies in this chapter and on the trends at the national level, their influence on environmental sustainability, and their influence on the experience of modernity.

This chapter has identified some key trends of how life and the experience of energy-modernity in rural areas in Southeast Asia have changed. The above key trends are summarised by Rigg et al. (2012) as three generalised propositions: '[1] a delocalization of life and living, reflected most obviously in heightened levels of mobility; [2] a dis-embedding of households and families as social and economic relations are stretched across space; and [3] a dissociation of the village-community as the village 'covenant' is frayed and interests diverge'. (p. 1,470). These propositions are reflected to a more or lesser extent in the four energy trajectories. This chapter adds four further propositions, specifically related to transitions in energy end-use appliances: the first is that access to energy, increased connectivity and integration in the monetary economy has resulted in more use of appliances, vehicles, and use of these technologies for longer periods of time. Second, the energy consumption practices of the younger generations, who grew up with access to electricity and other sources of energy, are very different from those of older generations. Third, people in the rural areas in Southeast Asia are increasingly exposed and linked to the world beyond their villages, through energy-based practices such as increased transport, communication and media: this draws out tensions between the

Table 5.5 Key end-use technologies from Chapter 5 and their influence on environmental sustainability and the experience of modernity

(National trends are included, subject to availability of data)

End-use technologies	Trends in case studies (based on survey sample)	Trends on national level	Influence on environmental sustainability	Other influences identified in chapter
Television	Nearly ubiquitous in Thai cases for almost two decades. Lao case studies have over 50% penetration rates now.	Percentage of television viewers in Thailand has increased from 80.4% in 1989 to 94.6% in 2008.[1] Percentage of households with television in Laos increased from 4.4% in 1994 to 30.3% in 2002.[2] No recent data available.	Increased electricity use, depending on size and usage, and on type of electricity production.	Changing worldviews, affecting children's education, challenging and reinforcing traditional culture, individualisation.
Mobile phone	Has become a common appliance since 2000s in all four case studies. Lao case studies have nearly same penetration rate, but fewer phones per capita.	Data very similar to case studies. In Laos, dramatic increase from 1 phone per 100 people in 2001 to 87 per 100 in 2011. In Thailand, in the same period from 12 to 113 per 100 people.[3]	Limited increased electricity use, and on type of electricity production.	Influence on experience of time-distance, influences travel behaviour, individualisation.
Motor bike	Saturation in Bo Nok and Mae Kampong and over 50% penetration in Lao case studies.	Data is limited and might not be reliable. In Laos, there were 67 'passenger cars' per 1000 people in 2010. Prior data unavailable.	Increased use of petrol. Biofuels still marginal and controversial.	Influencing experience of time-distance, influencing travel behaviour, individualisation.
Car	Less than 45% of households in Thai case studies. A few in Nam Ka, but none under respondents in Phakeo.	In 2010, there were 157 'motor vehicles' per 1000 people in Thailand, compared to 134 in 2006.[4]		stretching of geography of cost and benefit (petrol).

Tractor (and other agricultural mechanisation)	Ubiquitous in Bo Nok, including other machines. Rapid increase in Lao case studies with 25–50% penetration rate. N.a. for Mae Kampong	No data.	Increased use of petrol. Biofuels still marginal and controversial.	Changing farming practices, influencing social relations in the village (rent), used for off-grid electricity generation and transport, stretching of geography of cost and benefit (petrol).
Cooking – wood	Not much use in Bo Nok. Still important in Mae Kampong and ubiquitous in Lao cases.	Biomass accounted for 33% of the household energy consumption in 2000.[5] In 2011, 2.1% of the total energy expenditure went to biomass and charcoal and 3.3% to gas used in the household.[6]	Local environmental impacts (depending on forest and population density).	Health impacts, time consuming, provides warmth, gendered impact, used to prepare sticky rice (Laos, North and Northeast Thailand).
Cooking – charcoal	'Transition' fuel for Mae Kampong, sparely used in Bo Nok. Rarely used in Lao cases.	In Laos, approximately 80% of Lao households used wood as the main energy source for cooking, 15% used charcoal and 1% used electricity or gas in 2005.[7]	Local and regional environmental impacts.	Used for grilling, gendered impact.
Cooking – gas and electricity	Gas commonly used in Thai case studies, electricity sporadically. Both are not used in Lao case studies.		Long-distance environmental impact. Impact of electricity use depends on type of electricity production.	Changing food habits, changing experience of time-distance, gendered impact, energy security issue, stretching of geography of cost and benefit (gas, electricity).

[1] Data from NSO (2012).
[2] Data from NationMaster (2013).
[3] Data from World Bank (2013).
[4] Data from World Bank (2013).
[5] Data from IEA (2006).
[6] Data from NSO (2011).
[7] Data from Messerli et al. (2008).

Source: Author, unless indicated otherwise.

traditional cultural practices and those perceived to be more mainstream or modern. Finally, energy transitions are linked to important changes in the experience of modernity, such as a changing experience of distance and time, a shift in balance from place-based community to the individual, and an increasing gap and stretching of the geography of cost and benefit. The next chapter consolidates these findings, as well as those from the previous two empirical chapters.

Chapter 6
Towards a Critical Geography of Energy Transitions

Introduction

In this chapter, I discuss the key contributions of this book, focuses on energy transitions, modernity, sustainability and their interactions. The first part of this chapter elaborates on the key argument of this book, that is, energy and modernity are intimately related and cannot be separated. The second part of the chapter argues that sustainability needs to be seen as part of this energy-modernity dialectic – rather than opposed or outside of it – in order to move beyond superficial and discursive sustainability claims. The final part of the chapter broadens the scope to see what these propositions mean for a possible critical geography of energy, drawing on some of the key concepts that featured in this book.

Rethinking the Energy-Modernity Dialectic

Modernity as Diffused and Dispersed Condition

This book has challenged the conceptualisation of modernity as a distinctly different phase or period of time. One of the starting points of this book was that modernity can be conceptualised as a philosophical and epistemological condition, or as an historical and empirical instance (Wagner, 2001, p. 3). The former conceptualisation is often used in discourses of modernity of the state and other actors. The latter conceptualisation, which is used to understand modernity as a dynamic process of change, is a more reflective approach based on these discourses as well as on the lived experiences and aspirations of different actors. The empirical chapters show that the conceptualisation of modernity as distinctive historical or empirical instance, cannot be maintained, both for the country-level and local energy trajectories in Southeast Asia. Simple questions demonstrate this point. For example, did Thailand's modernity start in the mid-nineteenth century with the reforms introduced by King Mahidol and Chulalongkorn under pressure from the European colonial powers in the region? Or, did it start in the period after the Second World War (post-1945) with the start of the long period of rapid economic growth? Posing these questions on the level of the energy trajectories runs into similar problems. For example, did Nam Ka become a modern village after it was officially registered by the government following the Revolution? Was it when the villagers first got access to a road, or when their micro-hydropower system was built? Has Bo Nok 'missed out' on becoming a modern sub-district by not having a coal-fired power plant, as some of the supporters of this project suggest? The fact that there is no single answer to the above questions shows the pitfalls of conceptualising modernity as a distinctly different or historical phase.

However, discourses of modernity – which are often either implicitly or explicitly based on understanding modernity as an historical phase – have been shown to be important when analysing the development of energy transitions and modernity at different scales. The country-level case studies show that state-led discourses of modernity have been important drivers of energy sector developments at various moments in time, for example the territorialisation of the electricity grid and the establishment of state-owned utilities. Another example is the Lao government's goal of pursuing a 90 per cent household electrification rate by 2020, as part of its strategy to graduate from its least-developed country status. This state-led discourse can be found in many government documents and is widely quoted in reports from other actors – and academic papers – often without any official reference. Moreover, at the local level, access to electricity, and grid electrification in particular, is seen as one of the most important indicators of modernity for many people. These examples demonstrate that experience and discourses of, and the pathway to modernity, are neither singular nor unified. Rather, they show that modernity is dispersed, context-specific and diffused, with different meanings for different actors. Moreover, they stress the dynamic nature and mutual shaping of energy and modernity over time and in different contexts, and therefore resemble the notions of multiple or alternative modernities (Gaonkar, 2001; Knauft, 2002).

Such an understanding of energy-modernity begs the question of what keeps this dispersed and context-specific modernity together as an academic concept. This question may be answered through the analogy in a debate within the discipline of human geography about 'neoliberal natures', which faces a similar issue. The starting point of this debate is the fact that geographers increasingly understand neoliberalism beyond an ideal-type hegemonic idea, but rather as 'actually-existing neoliberalisms' through case studies. However, according to Noel Castree, this raises some important questions. For example: what holds neoliberalism together; or, how can we connect the dots between the many case studies? (Castree, 2005, 2006, 2008a, 2008b, 2009) The response to these questions is led by Karen Bakker (2009, 2010) who attempts to retain the idea of neoliberalism by 'develop[ing] conceptual frameworks that might account for variegation as a dialectic between geoinstitutional differentiation and translocal (but not generic) patterns and processes' (Bakker, 2010, p. 722). This book has tried to do just that, to develop a conceptual framework for the understanding of energy-modernity without having to throw out the concept of modernity with its analytical bathwater. Following Bakker's conceptualisation, this book rejects the idea of modernity as an ideal-type, coherent category, but rather sees it as a starting point for scalar and comparative analysis. Moreover, retaining the concept of modernity also helps to show how variegation is mediated and co-constituted in relation to energy and sustainability (Bakker, 2010).

One of the key advantages of the conceptualisation of energy-modernity as a diffused yet analytically useful concept is that it goes beyond the idea that the consequences of modernity are inherently positive or negative. As demonstrated in Chapter 2, this tends to be the case in the Ecological Modernisation Theory or political ecology literature. It also triggers one to go beyond the narratives that pitch economic development against environmental sustainability, as is often the case for large-scale power plant projects in Thailand and Laos. Finally, by focusing on analyses of energy trajectories and 'everyday' examples of the energy-modernity dialectic, this research has shown how the dynamic and varied nature of changing energy systems impacts on the lives of people.

Situating the Energy-Modernity Dialectic in Four Bodies of Literature

The notion of energy-modernity as a dispersed and non-monolithic philosophical condition sheds new light on some of the implicit or explicit conceptualisations of modernity emerging from the literature, and on the empirical findings of this book. Chapter 2 shows that the four bodies of social science literature – socio-technical transitions, Ecological Modernisation Theory, energy practices and political ecology – deal with the instances of modernity identified in this book in different ways. Examples are the changing experience of distance and time, the shift in balance from the community to the individual, and the increasing invisibility of energy production infrastructure for most people (cf. Table 2.2). In some social science approaches, these kinds of instances are reduced to an opaque monolithic category without explaining how or why these changes take place, let alone how they are related to energy trajectories and transitions. The socio-technical transitions literature implicitly conceptualises modernity as something that is part of the 'landscape' level; this includes aspects such as climate change, wars and other global events. However, by doing so, it may reduce questions about energy to technocratic problems, omitting a context-specific and historical framework for understanding social change. On the rare occasions when socio-technical transition scholars refer to a theory of modernity, it is usually a rather selective and simplistic adoption of the sociology of Beck and Giddens. In contrast, two other bodies of literature discussed in this book – Ecological Modernisation Theory and political ecology –

situate issues of modernity more central to their respective analyses. However, some authors within these bodies of literature reduce modernity to 'capitalist modernity' or even just capitalism. This way of conceptualising modernity not only misses some of the micro-dynamics analysed in this book, but also creates an unproductive dichotomy between modernity and sustainability.

The understanding of modernity emerging from the energy practices literature is more in accordance with the findings of this book. This literature's conceptualisation of modernity as the 'construction of normality' and focus on the different elements of practice are also in line with the scalar and dispersed understanding of energy-modernity postulated in this study. The analysis of energy trajectories in Chapter 4 and 5 shows how modernity is actively constructed in energy practices through the interplay of technical artefacts, existing energy routines, discourses of modernity and livelihood changes at the local level. Technological developments, such as the extension of roads and electricity infrastructure and the dissemination of televisions sets, mobile phones and motor bikes, have gone hand-in-hand with changing perceptions of time and distance, the convergence and tensions between majority and minority cultures, and the increasing invisibility of energy. However, the energy practices literature needs to be complemented by political ecology insights when analysing the discourses of modernity. This literature has been proven important to understanding the country-level cases as well as the local case studies in this book.

Situating Sustainability in the Energy-Modernity Dialectic

Having discussed the findings contingent to the discussions surrounding energy-modernity in the four bodies of literature, the unresolved question remains: where does sustainability fit into the energy-modernity dialectic? For governments and the private sector, and even for some NGOs, sustainability is often seen as something external or additional to their business-as-usual practices. This is similar to some of the bodies of literature reviewed in this book, in which sustainability is discussed in isolation from processes such as energy production and consumption, and therefore outside of the energy-modernity dialectic. This book, however, argues that sustainability should be seen as an inclusive part of this dialectic, rather than as something external to it. This argument is developed in more detail in the next section, which starts by discussing local energy trajectories, followed by country-level case studies, and then seeks to determine how the interactions and conflicting aspects of sustainability between these scales may be framed as 'competing sustainabilities'. In the final part of this section, I discuss how reflexive modernity helps to understand the relationship between energy, modernity and sustainability.

The Place of Sustainability in the Local Energy Trajectories

The argument in favour of the inclusion of sustainability into energy-modernity can best be made by looking at empirical examples in which sustainability is treated as outside of this dialectic. The decentralised generation systems in three of the cases discussed are all examples of where the idea of sustainable energy was implemented rather superficially, and disconnected – to a more or lesser extent – from the local livelihoods and contexts. In Phakeo in particular, there was a marked disconnect between the local livelihoods, the aspirations of the people, and the energy provided through their decentralised system. The French NGO Fondem insisted on trying out its model of sustainable energy through a hybrid solar-diesel village grid, without adapting it sufficiently to the local livelihoods and future energy demand. In Nam Ka, notions of off-grid – and later on-grid – decentralised generation were prioritised over the local energy needs, especially during the rehabilitation of the system in the public-private partnership. In contrast, Mae Kampong showed a more gradual evolvement of the system and energy-modernity, and more control and ownership by the local community. For these reasons, this may be considered a more successful case of integration of sustainable energy into the local energy-modernity trajectory. However, eventually the use of the micro-hydropower system in this place diminished in favour of grid electrification.

One of the key reasons underpinning the problems of integrating sustainability into the livelihoods and everyday practices of people in the three case studies – Phakeo, Nam Ka and Mae Kampong – was that the technical parameters of these socio-technical systems were often pre-determined and hardly open for negotiation, despite attempts to involve different stakeholders in the name of sustainability. In Phakeo and Nam Ka, for example, extensive surveys prior to implementation were undertaken and efforts were made towards participation and inclusion. In Nam Ka, the presence of Helvetas led to several 'participatory development' meetings and monitoring surveys. However, it is at best unclear how much influence these efforts had on the design of the system. The situation in Mae Kampong was, again, better in this sense because the community was more in control through the village cooperative which helped to better align decentralised electricity generation with the extant livelihoods. However, in the end, even in Mae Kampong the technical parameters did not prove flexible enough to accommodate the demand for increasingly stable electricity in the village.

The controversy over the siting of a coal-fired power plant in Bo Nok was an example of a site where the disconnect between energy-modernity and sustainability led to conflict, even bloodshed. In this case, the imminent environmental and social impacts of the project were resisted by a large part of the community: the resultant movement eventually prevented the coal-fired power station from being built. As such, the case of Bo Nok highlights the difficulty in finding alternative models of energy development. The community-owned projects, like the windmills, and the solar panels on the school, were largely symbolic and their impact on electricity production was marginal. The 1 MW private solar farm produced a more substantial amount of electricity, and the land could still be used for agriculture. But, there was little community involvement during the development of this particular project. Moreover, and unlike the other case studies, there was no direct benefit for people living in the area from the electricity generated by these decentralised projects because production and consumption of electricity in this area had long been separated. This particular issue is also key to understand the place of sustainability in Southeast Asia and in the country-level case studies of Thailand and Laos.

The Place of Sustainability in the Regional and National Level Case Studies

Chapter 3 follows the traces of a number of different discourses of environmental sustainability in Southeast Asia, discourses that have often been enmeshed with others including energy security and economic efficiency of energy production. Some examples are the discourses surrounding the oil crises of the 1970s, and those during the period after the revolution in Laos in 1975. Measures taken during these periods, such as the switch from small-scale diesel generators to hydropower, have left long-term effects on the structure of the energy sector and environmental sustainability. On the other hand, these kinds of 'crisis narratives' (Bridge, 2011) have been used to push for neoliberal energy reforms, such as the attempt to unbundle and privatise the Thai power sector in the late 1990s (Greacen and Greacen, 2004).

Hydropower presents an important example of the dynamic nature of energy-modernity, and the complexities of inclusion of environmental sustainability at the national and regional levels. Thailand started developing its hydropower potential in the 1950s. At that time, like many other countries in the global south, it followed in the footsteps of the United States, the Soviet Union (Russia) and the European countries in its attempt to develop hydropower as a clean and cheap source of energy (Smil, 1994).[1] As elaborated in Chapter 3, hydropower development in Thailand was part of a World Bank (1959) programme to fast-track Thailand into modernity, but sustainability was not an explicit concern. In contrast, hydropower dams in Laos were seen as the 'silver bullet' to achieve not only modernity and poverty eradication, but also clean energy and sustainability (Molle et al., 2009a). However, while the negative social

---

1   In 1954, Jawaharlal Nehru, who was prime minister of India at the time, referred to hydropower dams as the 'temples of modern India' (Roy, 1999).

and environmental impacts of hydropower have increasingly been emphasised by various actors, the economic benefits have often been overestimated. This line of critique culminated in an important report titled *Dams and Development: A New Framework for Decision-Making* by the World Commission on Dams (WCD, 2000) and the subsequent removal of (large) hydropower from the category of 'renewable energy' by many organisations. These examples show how thinking about sustainability has changed over time, sometimes challenging the prevailing energy-modernity dialectic, at other times reinforcing it.

The second point emerging out of the regional and country-level case studies is the complexity of scalar interactions and the resulting competing sustainabilities. This can be understood by analysing the different histories and trajectories within Southeast Asia. Within the borders of Thailand, for example, there seems to be an increasing awareness of – and protest against – the negative implications of large scale coal-fired, hydropower or nuclear power projects. As a result, many of these projects have been stopped, changed fuel-source or incurred serious delays, such as in Bo Nok. At the same time, there is a proliferation of alternative and renewable energy projects. Influential domestic stakeholders, such as Thai banks, are increasingly adopting corporate social responsibility commitments (Middleton, 2009). One of the consequences of these developments has been a shift towards the development of hydropower and coal-fired power stations in Laos, where there has been considerably less critical movement against state-led energy-modernity based on large-scale power plant projects. Moreover, Thailand's reliance on natural gas has led to the enrichment of the military regime in neighbouring Myanmar, which had hitherto not shared the benefits of the resource exploitation and has been embroiled in longstanding conflict with civilian populations in many parts of the country (Kolås, 2007). The construction of these large-scale energy projects involved Thai construction companies and banks, as well as other companies from the region, unhindered by their domestic corporate social responsibility commitments. As shown in Chapter 3, this issue of competing sustainabilities has been further compounded by assumptions about the economic efficiency and environmental sustainability of regional power grids in Southeast Asia. These regional discourses have largely hinged upon the availability of cheap and 'clean' hydropower and gas from Laos, Vietnam and Myanmar (Greacen and Palettu, 2007; Hirsch, 1998; Middleton et al., 2009).

Reflexive Energy-Modernities

Having identified the empirical ways in which sustainability is related to the energy-modernity dialectic, in this section I discuss reflexive modernity – as introduced in Chapter 2 – as a way to link the two concepts. For Beck and Giddens, 'simple' modernity based on Enlightenment principles has reached its limits. But, unlike post-modernist thinkers such as Lyotard and Baudrillard, this has neither seen them opposed to modernity nor resorting to ideas about alternative modernities (Beck, 1992; Beck et al., 2003; Beck et al., 1994). Rather, they propose the idea of reflexive modernity, also referred to as radicalised modernisation, as a way to analyse social change related to globalisation, individualisation and the proliferation of risk (Dodd, 1999). Reflexive modernity does not presuppose a certain direction or pathway of reflexivity; it is therefore different from 'classical' theories of modernity, such as Marx' theory of the evolution of economic systems or Rostow's stages of growth. Rather, reflexive modernity is based on the inherent open-endedness of processes such as globalisation, new social contracts and risk.

Some of the ways in which reflexive modernity has been taken up in the literature reviewed in Chapter 2 shows its clear limitations in light of the findings of this book. Most explicitly, the key authors of the Ecological Modernization Theory literature have used the idea of radicalised or reflexive modernity to argue for more – or super – industrialisation to overcome the problems created by 'classic' modernity. As argued earlier, this interpretation of reflexive modernity does not emphasise the radical open-endedness of reflexive modernity: it sticks to capitalist modernity as the key to resolving contemporary environmental crises. By contrast, many authors associated with political ecology are highly critical of capitalist modernity and would rather argue for an alternative or post-modernity as opposed to reflexive modernity. As critics within and outside these bodies of literature have pointed out, this stand-off has proven paralysing, unhelpful and reinforces the dichotomy between capitalism and sustainability (Forsyth, 2003; Gibson-Graham, 1996, 2008). Instead, Chapter 3 shows that the process of state-formation and the different state-led discourses of modernity, in combination with the spread and evolvement of capitalist relations and the wave of neoliberalisation since the 1980s, are all important to understanding the history of energy-modernity of Southeast Asia. Moreover, the inherent contradictions between development and local aspirations can only partly be explained to be the workings of capitalism (High, 2008). In other words, the essentialist link between capitalism and reflexive modernity, as often put forward by EMT and political ecology scholars, does not hold.

Besides rejecting capitalism as the main driver of modernity, this book also rejects the idea of a radical split between simple, first or classical modernity and second or reflexive modernity. Beck claims that reflexive or second modernity is a distinctly different phase, largely based on the changing form of industrialisation and society in Western Europe (Beck et al., 2003). This book shows the shortcomings of this position from the perspectives of countries in the global south, such as in Southeast Asia. There are many points in this book that demonstrate that the project of modernity is not finished: nor is it entering a second phase. An example is the continuous importance of the three energy utilities in Thailand – EGAT, MEA and PEA – which were set up during the modernisation era of the 1960s. While they no longer exercise absolute monopoly of power generation, transmission and distribution, they continue to wield power through political lobbying, the setting up of subsidiaries, and advertising and associating themselves with national symbols and the Thai royal family. Moreover, the large numbers of western consultancy and construction companies and banks active in the development of large hydropower and other infrastructural projects do reveal change, but not a radical break in modernity.

Theoretically, the break from simple to reflexive modernity has been challenged by Bruno Latour in his provocative essay *We Have Never Been Modern* (1993) as well as in his work directly reflecting on Beck's concept of reflexive modernity (Latour, 2003). As Latour contends: "'reflexive' does not signal an increase in mastery and consciousness, but only a heightened awareness that mastery is impossible and that control over actions is now seen as a complete modernist fiction. In second modernity, we become conscious that consciousness does not mean full control' (p. 36). In other words, while Latour appears critical of the idea of second modernity, he tries to retain the idea of reflexive modernity to show the open-ended and contingent connections between the different elements, scales and ambiguities that define the boundaries of domains such as society, nature and science. This form of reflexive modernity reveals the dispersed nature of agency in networks and assemblages, and the ambiguities of modernity.

Another way of employing the idea of reflexive modernity can be found in the work of authors associated with the socio-technical transitions body of literature. In their book titled *Reflexive Governance for Sustainable Development*, Voß and Kemp (2006), for example, argue that reflexive modernity not only means that modernity should reflect on the impacts created by 'classic' modernity, which they call first-order reflexivity; rather, they emphasise the need for second-order reflexivity, which 'has brought up critical reassessments of rational problem-solving methods and led to the development of alternative methods and processes of problem handling that are more open, experimental and learning oriented' (p. 6). This distinction helps to understand the stand-off between environmental sustainability and capitalism referred to in Chapter 2. This shows that many approaches dealing with sustainability only focus on first-order reflexive modernity, which is likely to produce only short-term and superficial solutions. This book provides a number of examples of first order reflexive modernity, such as the setting up of a demand side management unit in the energy utility EGAT while continuing to work under a 'cost-plus' model that incentivises the sale of more electricity rather than energy efficiency (Greacen and Greacen, 2004). The fact that this department was hosted by the Corporate Social Responsibility division underlines this example of first-order reflexivity. Other examples included the large amounts of resources that have gone into the development of experimental renewable energy solutions for rural areas in Laos – such as the examples of Nam Ka and Phakeo show – without sufficiently challenging the prevailing logic and discourses of modernity of EdL and its bilateral and multilateral funders. These kinds of solutions fall into the category of what Swyngedouw (2010) terms the 'post-political', or first-order reflexive modernity solutions that do not challenge the existing state-led discourses of energy-modernity because they exclude fundamental underlying questions from the realm of politics.[2]

To conclude, reflexive modernity – which recognises the open-endedness of modernisation and does not take the break with modernity as a given – is a useful concept for understanding the interplay of sustainability and energy-modernity. Its use avoids some of the pitfalls of falling into a conceptualisation of energy-modernity developments as singular, unified or naively normative processes. Keeping this in mind, this chapter now turns to some programmatic reflections on how to incorporate the scalar analysis of energy transitions, modernity and sustainability within the social sciences.

Towards a Critical Geography of Energy?

In this last section of the discussion chapter, I explore the outline of an approach that could be called 'a critical geography of energy'. The first part of this section excavates the origins of the 'geography of energy' and argues that more academic work under this heading is justified, in particular through renewed emphasis on energy, modernity and sustainability. Moreover, some of the key concepts used throughout this book – embeddedness, the geography of cost and benefit, competing sustainabilities and energy trajectories – can help to shape this field.

As touched upon in Chapter 1, the 'geography of energy' or 'energy geography' is not a new field. In his book review essay entitled 'The Geography of Energy and the Wealth of the World', Pasqualetti (2011) describes how from 1950 until around 2000 the focus of this type of literature was mainly on discoveries of fossil fuel deposits, facility siting, land use, nuclear power and risk assessment (see also Solomon et al., 2003). In the twenty-first century, the emphasis of energy geography has shifted towards issues such as climate change, energy poverty and social justice, energy security, renewable energy and

---

2 The case of Bo Nok, in which an assemblage of different actors and symbols challenged the state-led energy-modernity discourse, may be seen as an exception in this book.

urban energy environments.[3] More recent examples of recent developments in the field of the geography of energy include a seminar series on the 'Geographies of Energy Transition' convened in the UK (Bridge et al., 2013) and the establishment of an Energy Geographies Working Group at the Royal Geographical Society in 2011; and, for Southeast Asia, a special issue on 'Actors, Interests and Forces Shaping the Energyscape of the Mekong Region' (Kaisti and Käkönen, 2012).[4]

Despite the renewed interest in the geography of energy, there are still many dimensions of this concept that deserve more research. Some of these feature in this book. Moreover, while the review by Pasqualetti (2011) points to some publications that integrate social theory into the geography of energy, his essay is still largely limited to the study of macro-level perspectives and supply-side issues (see also Solomon et al., 2003). While there is probably a wide readership for publications that discuss and link climate change, global capitalism and fossil fuel markets, there is the risk that they may blend together as shallow explanations for the workings and failures of the global macro-economic system, omitting analyses of political economy and social justice issues. Moreover, such an approach might preclude potentially fruitful links with the micro-level perspectives explored in this book, including insights from anthropology, sociology, development studies, and political ecology.

Based on the above observations, the review of the literature and the empirical findings, this book argues that there is still considerable scope for a critical geography of energy. The discipline of geography is in a good position to contribute to a comprehensive issue-based framework, with a deeper scalar understanding of energy, modernity and sustainability. This resonates with the following comment by Hirsch (2012) on the role of geography in agrarian studies:

> [T]here is a both a logic and a demonstrable case to be made for the emergence of inherently geographical questions of space, scale, regional specificity and human environment relations ... moving on from some of the local anthropological approaches, on the one hand, and the broad sweeping political science questions, on the other. (p. 402)

More specifically, in relation to energy landscapes and geography, Bouzarovski (2009a) notes that:

---

3 Copied from a table derived from the analysis of 203 books (Pasqualetti, 2011, p. 975).

4 While this special issue was part of a development studies journal, many of the articles have a strong geography focus and draw broadly on a political ecology approach.

[w]ith their established knowledge of the multiple spatial outcomes resulting from the interactions of infrastructure, society and economy at different scales (for example, see Solomon et al. 2004), geographers are ideally positioned to take a leading role in this process, as the discipline already has the analytical tools to interrogate the role of spatially embedded path-dependencies and development trajectories in the energy restructuring process. (pp. 461–2)

More succinctly, Bridge et al. (2013) state that: 'the low-carbon energy transition is fundamentally a geographical process that involves reconfiguring current spatial patterns of economic and social activity' (p. 331).

The above emphasis on geography, however, does not mean that such an approach would be limited to this discipline only. This book has benefited, for example, from insights drawn from (environmental) sociology, anthropology, political economy and development studies. Therefore, a critical geography of energy should be relevant for these disciplines, as well as for other fields of social sciences, engineering and the sciences more generally. The rest of this section explores some of the building blocks of a critical geography of energy, as set out in Chapters 1 and 2, complemented with examples from the findings of this book.

The Embeddedness of Energy-Modernity

This book has made a case for understanding the energy transitions in Southeast Asia as embedded in peoples' livelihoods. The concept of embeddedness builds on the widely-cited notion of Granovetter (2002 [1985]) that economic actors are embedded in social relations. Importantly, Granovetter also uses this concept to argue against the distinction between pre-modern and modern economies (Carruthers, 2005). This book expands his notion by arguing that energy systems and transitions are also embedded in social relations, which follows directly from the conceptualisation of the energy-modernity dialectic in which modernity is shaping energy and vice versa. In addition, there are two other ways in which embeddedness emerges out of the empirical research findings. First, the discussion of state-led modernity and energy transitions in Chapter 3 shows that developments in the energy sector cannot be separated from the political and socio-economic developments and discourses at the country and regional level, and how deeply intertwined these elements are. This use of embeddedness of energy infrastructure is an important theme in (urban) political ecology, particularly in the work of Graham and Marvin (1994, 2001). Moreover, in an article on the 'ethnography of infrastructure', Star (1999) refers to embeddedness as a way to understand infrastructure, as 'sunk into and inside of other structures, social arrangements and technologies' (p. 381).

Second, this book has shown the embeddedness and mutual interconnectivities between energy technology development, local lived experiences and meanings of energy-modernity. In doing so, this book bridges a gap between the approaches based on livelihoods and those focusing mainly on energy technology. On the one hand, livelihoods-oriented approaches tend to underplay the importance of energy production and consumption, seeing it as more focused on 'traditional' development issues, such as agriculture, livestock, culture and water. The NGO's general lack of interest in energy issues in Laos is a case in point here (cf. Smits and Bush, 2010, p. 120). On the other hand, energy-technology focused approaches often do not take livelihoods and socio-economic contexts seriously, seeing them as part of the realm of engineers and economists (Lutzenhiser, 1994). A concrete example provided by this book is that one of the lead engineers involved in the setting up of the hybrid solar village grid in Phakeo was unaware that the village had only been established in 1998 in a resettlement area. Yet, as this book has demonstrated, this dynamic is crucial to both understanding the choice of this site, and how the new electricity generation system affected the livelihoods of the people in Phakeo.

An Expanded Geography of Cost and Benefit

This book has provided examples of how scale plays a role in the understanding of energy-modernity and sustainability, which a geography of energy could build upon. First of all, it has demonstrated that most forms of energy production and consumption involve a specific geography of cost and benefit, which is reconfigured in the process of transition. This not only goes beyond the supposition that all social and environmental 'costs' can be quantified and compared in monetary terms, but also beyond the dogma of the need for trade-offs between 'local' versus 'national' interests (Hirsch, 1998, 1999). The development of large hydropower dams – where the social and environmental costs are generally borne by the people at the dam site while the electricity flows to the urban areas – is one of the key examples here. This type of centralised electricity generation and transmission relies on the decoupling of the place of electricity production and consumption and often a scaling up, or 'stretching', of the geography of cost and benefit.

More specific examples of the changing geography of cost and benefit are found in the local case studies in this book. In Nam Ka, for example, the whole village helped to rehabilitate the micro-hydropower system; yet, only approximately half of the households got the benefit of electricity (although some managed to connect 'illegally' through their neighbours). The case of Bo Nok shows a different aspect of the geography of cost and benefit. The plans for the construction of the power plant were initially framed by actors supporting the power plant as a 'local' versus 'national' interest problem. In other words, local interests were subordinated to the objectives of industrialisation and economic growth. However, the power plant opposition network managed to turn this controversy into a discussion about national energy policy, and to expose the 'national interest' as (in their view) elite and business' interests.

This book also extends the geography of cost and benefit to the field of energy consumption. Hirsch (1998) mainly used the geography of cost and benefit to discuss the dynamic of energy production and its local impacts, while treating the (urban) consumption of energy largely as a black box. This book has unpacked this black box by exploring energy trajectories which involve both changes in production and consumption within a relatively small area. This has allowed for analysis of the development of energy consumption practices in parallel with the production of energy. However, this geography of cost and benefit changed more radically in Nam Ka and Mae Kampong when the decentralised systems were connected to the national grid. An important insight of these cases has been to show the implications of this re-scaled geography of cost and benefit for the increasing invisibility of energy infrastructure (Shove, 1997). Put simply, the further away the place of production is from the place of consumption – both physically and mentally – the more energy becomes an abstract issue which tends to be taken for granted. Meanwhile, the costs of the expansion of transmission lines are often heavily supported by soft loans and cross-subsidies, making it very hard for decentralised renewable energy sources to compete. The consumption side of the changing geography of cost and benefit helps to facilitate an understanding of why reducing energy consumption is difficult under these circumstances of normalising and escalating energy practices in increasingly urban or urbanised environments (Bulkeley et al., 2011; Shove, 2003a; Urry, 2011).

Scalar Politics and Competing Sustainabilities

The increasingly complex and rescaled geographies of cost and benefit reveal the importance of the politics of scale. These issues have gained widespread attention in the human geography and political ecology literature (Herod, 2009; Sheppard and McMaster, 2004; Swyngedouw and Heynen, 2003). However, this book links the concept of 'competing sustainabilities' with the politics of scale, which are relevant for a geographical approach to energy issues.

The politics of scale links with the politics of environmental sustainability in what might be called 'competing sustainabilities'. Rigg and Jerndal (1996) write about overlapping or competing sustainabilities as follows: '[W]e believe there to be multiple, overlapping sustainabilities, not one neat concept and reality. Indeed, it is the conflict which exists between different user groups and their different notions and practices of sustainability which gives rise to many of the problems' (pp. 141–2).

An example of competing sustainabilities drawn from this book is the emphasis on hydropower as 'clean energy' by the government of Laos and in the regional power development plans of the Greater Mekong Subregion of the ADB and ASEAN. However, for a country like Laos, and especially for people living in dam-affected areas, there are clearly negative implications for environmental sustainability as well as for their livelihoods.

Another example is the 'success' of Mae Kampong as an eco-tourism community. While this form of tourism has become a key source of income for the village cooperative, it has also put increasing pressure on the land and forest resources, waste management, social cohesion of the village, and energy consumption. Moreover, the success of sustainable tourism has not managed to halt the out-migration of young people to the city. Not many jobs were created and those that were, were often filled by people from other areas. But, these issues are conveniently glossed over in the mainstream discourses of Mae Kampong as an iconic model of sustainable 'alternative' energy and community development.

Climate change is perhaps the most important example of a discourse heavily implicated in the politics of scale and competing sustainabilities. While this book does not directly engage with the impact and policies of climate change, it nonetheless features in many of the case studies. People in Mae Kampong, for example, routinely refer to the issue of climate change to defend their community micro-hydropower project and eco-tourism activities. For the local activists in Bo Nok, it is an important way to mobilise NGOs such as Greenpeace to support their campaign. Meanwhile, utilities such as EGAT are trying to show their efforts to reduce pollution and promote renewable energy. Hydropower, again, is one of the technologies that has experienced revival due to its perceived status as a clean and renewable energy source.

The above examples show that sustainability is contested and context-dependent, despite attempts to incorporate it in discourses that favour sustainability in one scale over another. A focus on national plans for sustainability, for example, might not be in line with global sustainability targets; similarly, these plans might not be able to account for the sustainability of local energy trajectories.

From Energy Transitions to Trajectories

A critical geography of energy should be careful in using the concept 'energy transition', which has some clear limitations compared to 'energy trajectories'. Whereas the former is a well-established concept and plays a key role in academic literature, the latter is a neologism that has emerged during the process of developing this book. This section argues that the idea of energy trajectories seems better equipped to deal with the embedded, multiple, contingent, networked nature of the energy-modernity dialectic.

There are a number of similarities between energy transitions and trajectories: both can be used to analyse developments related to energy over a period of time and within a certain spatial context. Chapter 2 has shown that the term 'energy transition' has various definitions and can be used in a number ways. Some authors narrowly focus on technical aspects of energy transitions, while others relate it to more systemic changes. There is no extant innate spatial context or temporal frame for either of these two concepts, although the nation-state often features in analyses of energy transitions (Raven et al., 2012). In contrast, the idea of energy trajectories was introduced to find a way to conceptualise energy transitions at the local level. While it would have been possible to use something like 'local energy transitions', the word 'trajectory' was introduced to stress the embeddedness of the changing energy systems, the multiplicity, as well as the contingency. The first aspect, embeddedness, has already been covered in this discussion and could probably be 'built into' energy transitions as some more systemic definitions have already done. The other two aspects will be discussed in turn below.

The multiplicity of energy trajectories, however, is an important point of difference between transitions and trajectories. Each of the local case studies (trajectories) includes multiple sources of energy as well as different energy practices. For example, the introduction of electricity has not automatically led to the phasing out of the use of firewood for lighting and cooking, nor does it account for the different energy needs of other livelihood aspects, for example communication, entertainment, transport and agriculture. The idea of energy transitions suggests uniformity or at least a convergence when it comes to these different energy uses. As such, the concept of energy transition bears some of the hallmarks of 'classic' or simple modernity, namely the progressive or stepwise 'improvement' in the use of energy, whereas trajectories does not.

Contingency is another key element of energy trajectories. For each of trajectories discussed in this book, the impacts of the changing energy systems over time have led to various changes in people's energy practices, worldviews and cultures, sometimes in unpredictable ways. Moreover, these changes have not been uniform: their impact on different groups, for example on the rich and the poor, the young and the old, and the ethnic minorities and the majority has been varied. These developments do not just depend on an understanding of the mechanisms involved, but also on the specific context and interplay between different elements. To this end, the trajectories show a strong degree of contingency and non-linearity. The concept of energy transitions, on the other hand, leaves little room for such contingency, but rather suggests a certain unidirectionality, and end state or teleology.

A final point of discussion is the possible difference in the scale on which energy transitions and energy trajectories operate. While the term 'energy trajectories' was introduced in reference to local case studies, energy transitions do not necessarily operate at a 'higher scale' compared to energy trajectories. In order to understand the complexities surrounding to energy-modernity and sustainability at different scales, the concept of energy trajectories is more coherent with concepts discussed in the previous section, for example the geography of cost and benefit, politics of scale and competing sustainabilities.

Normative Positions and Policy Recommendations

The final argument in this section discusses the place of normativity in a critical geography of energy, drawing on the main bodies of literature used in this book. While some of the discussions in the literature stress the need to be careful when taking a normative position on aspects of modernity and sustainability, total avoidance to take a position is not possible. Some words of warning on this topic may be taken from the debate between socio-technical transition and energy practice scholars about 'transitions management' and its normative aspects (Rotmans and Kemp, 2008; Shove and Walker, 2007). The debate centres on the experiment in Dutch energy sector to manage energy transitions towards sustainability. Adding to the well-documented outcomes and criticisms of this experiment, Shove and Walker (2007) raise a more general question vis-à-vis normativity and energy-modernity:

> In a manifestly complex world dominated by hegemonic ideologies of neoliberal capitalism, global finance, and commodity flows is it really possible to intervene and deliberately shift technologies, practices, and social arrangements – not to mention their systemic interaction and interdependencies – onto an altogether different, altogether more sustainable track? (p. 763)

This question shows that it is not only hard to analyse energy-modernity due to its complexities and interrelations, but even harder to employ a deliberate policy or management approach.

As well, there are normative positions to be found in some interpretations of reflexive modernity. Dodd (1999) argues that the demise of modernity as a teleological project, and the rise of the risk society have prompted Beck and Giddens to argue for a new kind of politics from below, which can mould our futures in new directions. More importantly, Dodd argues, they do not appeal to deeper moral or theoretical foundations, but claim that this state of reflexive modernity is already upon us. A clearer way of pinpointing the normative implications of reflexive modernity may be found in Lash's contribution in the co-authored book on *Reflexive Modernization* (Beck et al., 1994), wherein he discusses what reflexive modernity means in practice:

> [from] vertically and horizontally integrated, functionally departmentalized firms ... [to] flexible disintegration into networked relatively autonomous knowledge-intensive firms ... [From] totalizing inversion of the social rights of the Enlightenment project ... [to] welfare services that are a client-centred coproduction and decentralised citizen-empowered alternative ... [From the] abstract 'blue-print Marxism' of the Eastern European past and Western combination of capitalist state bureaucracy and abstract procedural parliamentarism ... [to the] politics of radical, plural democracy, rooted in localism and the post-material interests of the new social movements. (Lash, 1994, p. 113)

The above quote from Lash reveals a somewhat utopian interpretation of reflexive modernity. However, as this book has shown, some actors employ discourses and technologies such as decentralisation, localism and citizen-empowerment without achieving long-term sustainable practices. Their failure to sufficiently integrate these ideas and technologies into existing livelihoods risks perpetuating existing power relations, inequalities or forms of corruption. Moreover, these actors are sometimes, either willingly or unwillingly, caught up in state-led discourses of energy-modernity.

Political ecology scholars on occasion adopt a normative stance, drawing upon ideas pertinent to social and environmental justice. However, as Blaikie (2012) argues in his article titled 'Should some political ecology be useful?', the rise of post-structuralism and post-development has led to some hesitation when it comes to engaging with actors and stakeholders within the fieldwork area and outside of the academy. Notwithstanding, he argues for an engaged political ecology, while at the same time being aware of some of the dangers and opportunities related to:

> [on the one side,] co-optation, post-colonial rip-off, reinforcement of centralised control and the reproduction of environmental injustice. On the other side, there are charges of irrelevance, coffee table talk, clever words, introversion, academic promotion but no action outside the Ivory Tower and a denial of social responsibility. (p. 238)

Among other key contributions of political ecology are the scrutiny of the roles of science and academia in shaping debate appertaining to human-environment relations, drawing on insights from the Sociology of Scientific Knowledge and Science and Technology Studies, notably Actor-Network Theory (ANT). This theme has evolved from critical engagement with simplistic explanations of environmental degradation (Blaikie and Brookfield, 1987) to the study of global environmental discourses (Adger et al., 2001) and, finally, to challenging the role and position of science itself (Forsyth, 2003; Stott and Sullivan, 2000). This latter theme questions whether 'science' is inherently more legitimate than other forms of knowledge. Recent examples of such critical approaches in the field of energy are the ways in which the International Energy Agency is pre-empting more radically different energy scenarios (Labban, 2012) and an assessment of the 'radically conservative vision' of the United Nations Environmental Programme (UNEP) in its programme 'Towards a Green Economy' (Brockington, 2012).

A critical geography of energy should draw upon the insights from each of the above theoretical discussions about normative positions. The empirical findings in this book provide some practical insights into how a critical geography of energy could be used or abused. First, this book has made a case against the introduction of technological systems without a clear understanding of the livelihoods, actors and wider context. While in many places the use of small-scale renewable energy technology may be a good idea, the examples of Nam Ka and Phakeo provide cautionary examples of negative implications, in particular when the social part of the socio-technical system is undervalued. In other words, small is not always beautiful (cf. Schumacher, 1975) or profitable (cf. Lovins, 2002). Second, care should be taken when interpreting figures and statistics relevant to macro-level energy transitions and, in particular, when contemplating how to steer or manage them. This follows not only from the discussion about transition management in the Netherlands, but also from the empirical and theoretical unpacking of the concept of energy transitions itself. The concept of energy trajectories is better placed to capture the inherent reflexivity, embeddedness and ambiguity of energy-modernity in specific contexts. Notwithstanding the previous two points, avoiding taking a normative position is not an option. As authors in critical political ecology and STS have demonstrated, academic or scientific research does not take place in a vacuum, and science does not hold a privileged position. Finally, this book reflects a certain implicit and explicit positionality, through its selection of issues, literature, case studies and data, as well as the approach and epistemology adopted. This shows that a critical geography of energy has no choice but to become an engaged discipline.

Concluding the Discussion

This chapter has taken the key concepts of this book head on, drawing both on the academic debates in Chapter 2 and the findings from the empirical chapters. It has put some flesh on the bones of the energy-modernity dialectic by framing it as a diffuse and dispersed process, avoiding relativism but retaining the concept of modernity. Reflexive modernity, which helps to situate sustainability within this energy-modernity rather than outside of it, stresses that sustainability cannot be separated from the mutual shaping of energy and modernity. These key theoretical contributions led to some programmatic notes on a critical geography of energy. Some of the key concepts and implications for such a geography of energy include: an expanded geography of cost and benefit; the politics of scale and competing sustainabilities; and, the introduction of energy trajectories as a possible replacement for the term energy transitions. Finally, this chapter has made some comments about the need for a critical geography of energy to be cautious regarding – but not avoid taking – a normative position. The next chapter consolidates these and other key findings from the book and discusses their implications for theory and policy.

Chapter 7
Conclusion

Between Modernity and Sustainability?

The starting point of this book was to analyse the apparent tensions between modernity and sustainability in energy transitions at different scales in Southeast Asia, with a specific focus on Thailand and Laos. These neighbouring countries were chosen not solely because they represent very different political systems and levels of economic development within Southeast Asia, but also because of their shared language and culture and increasingly mutually dependent energy relations, in particular in the spheres of electricity trade and investment. Moreover, both countries are undergoing rapid changes in the production and consumption of electricity in tandem with major changes in urban and rural livelihoods. Finally, they reflect the fact that discourses about regional integration and cooperation are becoming increasingly important in Southeast Asia. This dynamic context provided the background for the following fundamental research problem: how to conceptualise the tension between the increasing need for energy and the environmental impacts and pressures posed by this need?

The key argument of this book is that energy and modernity cannot be separated, but are interlinked as an energy-modernity dialectic. In other words, the changes in modernity and changes in the production and consumption of energy, are mutually dependent and co-evolving. Moreover, the challenges faced when achieving environmental sustainability – for example, reducing per capita energy consumption and decreasing the social and environmental impacts of energy production and consumption – are an integral part of this dialectic. This book has reviewed the different bodies of social science literature which attempt to conceptualise the relations between energy, modernity and sustainability: socio-technical transitions, Ecological Modernisation Theory, energy practices, and political ecology. This review has in turn delineated the different conceptualisations of modernity and sustainability between and within these bodies of literature. This is not just an academic exercise: I want to stress that different approaches lead to different policy implications in the field of energy, environment and social change. Moreover, the literature review and empirical parts of this book have challenged some of these conceptualisations in order to construct a scalar approach to the study of the energy-modernity dialectic and the place of sustainability. In particular, the book argues for the centrality of the question of modernity, and against reducing modernity to the questions of capitalism or to a peripheral element.

This book has used different case studies to understand the relations between energy-modernity and sustainability at different scales. First, the processes and discourses of regional integration in Southeast Asia were analysed, with a focus on ASEAN and the GMS project. Next, the country-level case studies of energy transitions and state-led discourses of modernity in Thailand and Laos demonstrated the deeply intertwined history of energy transitions, state formation and discourses of modernity, drew comparisons where appropriate, and provided the background for the remainder of the book. This was followed by in-depth local case studies of the respective villages' energy trajectories, which involved not only transitions to different energy production systems and end-use energy technologies, but also their relations with the local livelihoods and changing energy practices. These cases aimed to provide a more fine-grained analysis of the workings of energy transitions and trajectories in Southeast Asia, and to challenge the implicit or explicit model of energy transitions as dominating local trajectories. A combination of qualitative and quantitative data at different scales was used to (re)construct the energy trajectories and ascertain how local energy practices have changed.

Key Issues and Arguments Linking Energy, Modernity and Sustainability

The first key argument was that developments in the energy sector and modernity in Southeast Asia are closely linked. Modernity on the regional and country-level is operationalised as the history of state-formation, territorialisation, state-led discourses of energy-modernity, and increasingly neoliberal discourses of regional integration. However, these macro-level changes cannot be understood without first looking at the changes in the everyday lives of people, their changing livelihoods, access to media, appliances and infrastructure. The increasing importance of discourses of sustainability in the energy sector in Southeast Asian countries is often part of energy security and energy efficiency programmes. It may, therefore, be partly understood as the outcome of the historical processes and discourses of state-led modernity, and partly as an alternative to them. Examples of these alternatives in Thailand include: support for small-scale decentralised power producers, feed-in tariffs, energy efficiency programmes, but also the ascendency of nuclear energy. In Laos, however, there is less space for alternative discourses of energy-modernity due to the limited space for civil society in this country. These differences, and the multiplicity and scalar interactions of these configurations of energy, modernity and sustainability, show the limitations of universalistic models of energy transitions and linear models of modernity itself.

This book has closely scrutinised energy trajectories in four cases, to determine who were involved, how people experience them, how they are embedded in the peoples' changing livelihoods, and how they shape energy practices. My analysis of the four energy trajectories provides a nuanced understanding that sometimes diverges from those at the national and regional level. In all four case studies, data on electricity use and the uptake of appliances such as TVs and mobile phones showed a rapid change in energy practices over the last three decades. Moreover, the responses to the in-depth qualitative interviews confirmed how these changes have co-evolved with rapidly changing and diversifying livelihoods and increased monetary income, albeit often with increasing inequality as each of the case studies revealed.

My empirical analysis of the energy trajectories revealed a range of changing energy practices, such as cooking with gas instead of firewood, new forms of communication and use of agricultural machinery instead of manual labour or buffaloes. While some of these transitions were similar to the energy transitions at the country level, others were not, suggesting that they cannot simply be seen as an extension of state-led modernity or state-led energy development. Moreover, the energy trajectories analysed in this book bring out changing experiences of modernity, such as the perception of distance and time, increasing individualisation, and the increasing invisibility of energy infrastructure. The energy practices literature provides an understanding of how practices at the local level change as a result of these elements and become 'normalised' in hitherto isolated areas. The sum of these aspects ultimately affects the intensity of energy use and environmental sustainability. Yet, it also shows the difficulty of analysing sustainability outside of this complex web of socio-technical relations and scalar interactions, that is, outside the energy-modernity dialectic.

Both the country-level and local energy trajectories are embedded in changing livelihoods, aspirations and discourses of modernity. This has important repercussions for the place of sustainability within the energy-modernity dialectic. While some of the cases are seemingly about sustainable energy, they cannot be discussed in isolation of issues such as local ownership and participation, changing geographies of equality, territorialisation of the uplands, and the will to improve (Li, 2007b). Environmental sustainability, I suggested, is a subset of these interactions rather than separate or alternative from the energy-modernity dialectic.

Implications for Social Science Research

A first contribution of this book is the advancement of an understanding of modernity as the outcome of complex scalar interactions. It stresses the importance of discourses of modernity both for stakeholders in the energy sector as well as in the construction of peoples' everyday practices. This does not mean, of course, that everyone subscribes to a similar notion of modernity. Rather, discourses of modernity, whether proposed by the state or by other stakeholders, are heavily contested and ever changing due to global and local interactions. In that sense, it is appropriate to speak of multiple or alternative modernities, since they are a result of context-specific, yet related, activities at different scales. While this book has emphasised the links between energy and modernity, they also apply to other issues and resources; for example, water and land.

A specific understanding of reflexive modernity is another important contribution made by this book. While this study is broadly sympathetic to the ideas of Beck and Giddens, it does not agree with their conceptualisation of reflexive modernity as a radical break with 'first' or 'simple modernity'. Instead, this book has argued that there is no such break. The findings from this case study demonstrate the dynamic, plural and dispersed nature of modernity as well as its inertia. For example, there is an increasing awareness of the social and environmental impacts of large scale energy projects, and the energy landscape of Thailand, for example, is slowly incorporating more renewable energy and energy efficiency compared to a few decades before. At the same time, the institutions associated with state-led modernity are only slowly changing as a result of these developments, as they are firmly locked in by existing infrastructure and powerful organisations. The implication for the social sciences is to recognise the central importance of these energy-modernity dynamics.

The notion of reflexivity supports the proposition that sustainability should be located inside the energy-modernity dialectic, rather than external to it. This not only challenges the idea of 'sustainability transitions' as a different category from other transitions that are unfolding, but also means that the ideas about, and dynamics of, sustainability need to be explored and located at household, village, district, province, country and regional levels rather than restricted to one of them only. Consequently, another contribution of this book is its move towards a more nuanced understanding of scale and scalar interactions in the study of energy problems and the social sciences more generally. The key implication of this is that energy transitions in Thailand and Laos can neither be reduced to macro-level statistics, nor to its many energy trajectories. Moreover, the connection between energy and modernity means that analyses of sustainability should include both production and consumption, rather than only focusing on the energy supply-side.

Finally, this book advocates studying the above dynamics by adopting a reinvigorated multi-disciplinary approach to the geography of energy. Geography is well placed to study the complex scalar interactions related to energy, modernity and sustainability. This book has also argued for the inclusion of insights from other social sciences literature such as energy practices and political ecology literature, and selected insights from socio-technical transitions and Ecological Modernisation Theory literature. These insights need refinement and possible inclusion of additional theories and methodologies that can account for both the material and discursive aspects of changing energy-modernities. Notwithstanding – and since this book has argued for an engaged approach – the next section outlines some general policy implications and recommendations.

Policy Implications

The strong links between energy and modernity means that energy policy is about more than just energy. It also deals with livelihood opportunities, (re)distribution of wealth, and environmental implications, to name just a few. For this reason, energy policy needs to be (re-)politicised at all scales. It should not be the sole responsibility of engineers and economists, but instead be conceived moves to include civil society representation in the energy policies of Southeast Asia, there is much more to be done.

A further policy implication is that renewable and distributed energy policies need to be appropriate and aligned with the local context and livelihoods rather than imposed by state-actors (such as the coal-fired power plant in Bo Nok), NGOs (such as Fondem in the case of Phakeo), or private companies (such as Sunlabob in Nam Ka). Each of the cases in this book shows the difficulties encountered when connecting the power generation activities within the wider livelihoods and aspirations of the people living in these areas. A positive example of promoting the use of distributed (renewable) energy is Thailand's VSPP policy, which has led to many small-scale biomass, solar and other renewable energy projects. An example from outside of the region (and this book) is the German *energiewende* (energy turn) (Hillebrand, 2013), which includes a concentrated effort by the state, private sector, civil society and communities to work towards the evolution of more sustainable energy production and the phasing out of nuclear power.

In addition, my analysis of the extended geography of cost and benefit in this book shows what a reflexive energy-modernity might look like as a policy guideline. Its aim would be to reduce the physical and mental distance between the place of energy production and consumption as much as possible. This could be effected using technologies such as solar panels on rooftops, small-scale (biomass) co-generation plants and heat pumps, and technologies that use renewable energy sources and save the costs of electricity transmission. Most importantly, they would close the gap between the place of production and the place of consumption and bring the positive and negative aspects of producing energy closer to peoples' homes or workplaces. Reducing this gap – either physically or mentally – could lead to more sustainable energy practices.

Finally, this book underscores the importance of energy efficiency as a major opportunity in policy discussions in Southeast Asia (cf. Foran et al., 2010). In Thailand, energy efficiency has only recently (in 2012) been taken seriously and integrated into the Power Development Plan – undoubtedly an important first step. In Laos, the ADB has been involved in energy efficiency programmes, albeit with little result. This last example is a reminder that energy efficiency cannot be tackled solely through regional or national policy. An approach inspired by the energy practices literature would help to identify the practices related to energy (in)efficiency and to target the elements that constitute these practices. When examining cooling practices, for example, the elements may include any or a combination of materials (air-conditioners, construction), competences (the way we dress and manage doors and windows) and meanings (our ideas of comfort, or the 'ideal temperature'). In other words, these policy interventions would require more than 'orthodox' rational-economic behaviour change; in addition, they would target infrastructure, skills and cultures, and consider relations with other practices such as heating, working and cooking (Shove et al., 2012).

Research Agenda and Conclusion

Because this book covers a lot of ground, of necessity it is open-ended in its explanatory strength. This is not so much a limitation as it is an acknowledgement that all research is necessarily partial, not least science-based research. As such, this book echoes the call for a social science that recognises the importance of linkages, interactions and the deep connectivity between humans and their environments (Law, 2004).

An avenue for further research is to go beyond the nation state in the study of energy and social change, which is a common criticism of socio-technical transitions and Ecological Modernisation Theory literature. This book argues that energy transitions are made up of countless trajectories at different scales, all of which are contributing to or contesting energy transitions commonly 'measured' at the national level. Such statistics do not capture the transmission lines passing over the unelectrified villages near Phakeo, people protesting against the development of small biomass power generators in Prachuab Khiri Khan, and/or the six-year postponement of Thailand's nuclear plans subsequent to the Fukushima disaster in Japan. Rather, such examples can only be understood through in-depth case studies, which draw links and connections with other scales. Focusing too much or only on the nation state and the 'national interest' (which this book does to some extent in Chapter 3) hides the fact that energy trajectories and modernity are simultaneously deeply localised and globalised (Swyngedouw, 1997), and lead to disparate outcomes depending on background, class, livelihood and geographical location.

Furthermore, a critical energy geography would benefit from more analyses of the approaches already included in this book and from some of those left out. Omitted from this book are macro- and long-term approaches to the study of energy transitions and social change, such as the Viennese socio-metabolic transitions approach (Fischer-Kowalski, 2011; Fischer-Kowalski and Rotmans, 2009). Certain 'branches' of the literature discussed have also been excluded, for example the technological innovation systems (TIS) approach, which has some overlap with the socio-technical transitions literature (Geels et al., 2008; Markard and Truffer, 2008), and the social-constructivist ecological modernisation approach (Dryzek, 2005; Hajer, 1995). Finally, more engagement with Foucault's governmentality theory could contribute to the capturing of relations between discourses of modernity by the state, private sector and multilateral actors, and the shaping of peoples' identities.

To conclude, in this book I have demonstrated the complexities and scalar interactions of energy, modernity and sustainability, using the examples of four energy trajectories, and by analysing the discourses of energy-modernity in Southeast Asia, with a specific focus on Thailand and Laos. By examining the examples of the lived experiences and discourses of modernity at these various scales, the book has shown why achieving sustainable energy systems is much more than merely a technological question or a matter of 'more' modernisation. Instead, it shows that energy questions go straight to the heart of how we govern and practice our daily lives, institutions, countries and regions, whether in Southeast Asia or elsewhere. The importance of these issues will only increase as the population continues to grow, resources become scarcer and the impacts of climate change more severe and frequent. This does not mean that the future is only grim. Indeed, this book has shown a number of examples of reflexive modernity and the emergence of alternative energy-modernity discourses at the local, national and regional level. These demonstrate that more environmentally sustainable and socially inclusive energy futures are possible when involved actors challenge the discourse of energy-modernity often taken for granted.

Bibliography

ADB, 2000. *Technical Assistance for Regional Indicative Master Plan on Power Interconnection in the Greater Mekong Subregion.* Manila: Asian Development Bank and the Government of Norway.

———, 2005. *GMS Flagship Initiative: Regional Power Interconnection and Power Trade Arrangements – Summary.* Manila: Asian Development Bank.

———, 2009. *Building a Sustainable Energy Future: The Greater Mekong Subregion.* Manila: Asian Development Bank, p. 226.

———, 2010. *Update of the GMS Regional Master Plan.* Manila: Asian Development Bank.

———, 2011. Greater Mekong Subregion. Manila: Asian Development Bank. [online] Available at: http://www.adb.org/gms/ [Accessed 24 October 2011].

———, 2012a. *Asian Development Bank – Statistical Database System Online.* Asian Development Bank. [online] Available at: https://sdbs.adb.org/sdbs/index.jsp [Accessed 27 march 2015].

———, 2012b. *Key Indicators for Asia and the Pacific 2012.* 43rd edition. Manila: Asian Development Bank.

———, 2014. *Summary of Proceedings: 16th Meeting of the GMS Regional Power Trade Coordination Committee (RPTCC-16).* Seam Reap, Cambodia: ADB, GMS Regional Power Trade Coordination Committee.

Adger, W., Benjaminsen, T., Brown, K., Svarstad, H., 2001. Advancing a Political Ecology of Global Environmental Discourses. *Development and Change*, 32 (4), 681–715.

Andamon, M.M.D., Williamson, T.J., Soebarto, V.I., 2006. *Perceptions and Expectations of Thermal Comfort in the Philippines.* Proceedings of conference: Comfort and Energy Use in Buildings – Getting them Right. 27–30 April. Cumberland Lodge, Windsor, UK.

ASEAN, 1980. *The First Meeting of ASEAN Economic Ministers on Energy Cooperation, Bali, Indonesia.* Jakarta: Directorate General of Power, Ministry of Mines and Energy.

———, 1999. *ASEAN Plan of Action for Energy Cooperation (APEAC) 1999–2004.* Manila: ASEAN Ministers on Energy Meeting.

———, 2004. *ASEAN Plan of Action for Energy Cooperation (APEAC) 2004–2009.* Manila: ASEAN Ministers on Energy Meeting.

———, 2008. *ASEAN Economic Community Blueprint.* Jakarta: Association of Southeast Asian Nations.

———, 2010. *ASEAN Plan of Action for Energy Cooperation (APEAC) 2010–2015*. Manila: ASEAN Ministers on Energy Meeting.

Asia Times Online, 2011. Laos Takes Stock Step. *Asia Times Online*, 19 January. Hong Kong.

Ausgrid, 2012. *Ausgrid 2011–12 Summary Community Electricity Report*. Sydney: Ausgrid.

Baird, I.G., 2011. Turning Land into Capital, Turning People into Labour: Primitive Accumulation and the Arrival of Large-Scale Economic Land Concessions in the Lao People's Democratic Republic. *New Proposals: Journal of Marxism and Interdisciplinary Inquiry*, 5 (1), 10–26.

Baird, I.G., Barney, K., Vandergeest, P., Shoemaker, B., 2009. Reading Too Much Into Aspirations: More Explorations of the Space between Coerced and Voluntary Resettlement in Laos. *Critical Asian Studies*, 41 (4), 605–14.

Baird, I.G., Shoemaker, B., 2007. Unsettling Experiences: Internal Resettlement and International Aid Agencies in Laos. *Development and Change*, 38 (5), 865–88.

Baker, C., Phongpaichit, P., 2014. *A History of Thailand*. 3rd edition. Port Melbourne: Cambridge University Press.

Baker, S., 2007. Sustainable Development as Symbolic Commitment: Declaratory Politics and the Seductive Appeal of Ecological Modernisation in the European Union. *Environmental Politics*, 16 (2), 297–317.

Bakker, K., 1999. The Politics of Hydropower: Developing the Mekong. *Political Geography*, 18 (2), 209–32.

———, 2009. Neoliberal Nature, Ecological Fixes, and the Pitfalls of Comparative Research. *Environment and Planning A*, 41 (8), 1,781–7.

———, 2010. The Limits of 'Neoliberal Natures': Debating Green Neoliberalism. *Progress in Human Geography*, 34 (6), 715–35.

Bambawale, M.J., D'Agostino, A.L., Sovacool, B.K., 2010. Realizing Rural Electrification in Southeast Asia: Lessons from Laos. *Energy for Sustainable Development*, 15 (1), 41–8

Barney, K., 2007. *Power, Progress and Impoverishment: Plantations, Hydropower, Ecological Change and Community Transformation in Hinboun District, Lao PDR, a Field Report*. Toronto: York Center for Asian Research, York University.

———, 2009. Laos and the Making of a 'Relational' Resource Frontier. *The Geographical Journal*, 175 (2), 146–59.

Beck, U., 1992. Risk Society: Towards a New Modernity. London: Sage.

Beck, U., Bonss, W., Lau, C., 2003. The Theory of Reflexive Modernization. *Theory, Culture & Society*, 20 (2), 1–33.

Beck, U., Giddens, A., Lash, S. (Eds), 1994. *Reflexive Modernization: Politics, Tradition and Aesthetics in the Modern Social Order*. Cambridge: Polity Press.

Berkhout, F., Angel, D.P., Wieczorek, A.J., 2009. Sustainability Transitions in Developing Asia: Are Alternative Development Pathways Likely? *Technological Forecasting and Social Change*, 76 (2), 215–17.

Berkhout, F., Verbong, G., Wieczorek, A.J., Raven, R., Lebel, L., Bai, X., 2010. Sustainability Experiments in Asia: Innovations Shaping Alternative Development Pathways? *Environmental Science & Policy*, 13 (4), 261–71.

Biggart, N., Lutzenhiser, L., 2007. Economic Sociology and the Social Problem of Energy Inefficiency. *American Behavioral Scientist*, 50 (8), 1,070–87.

Bijker, W.E., Hughes, T.P., Pinch, T.J., 1987. *The Social Construction of Technological Systems: New Directions in the Sociology and History of Technology*. Cambridge, MA: MIT Press.

Bijker, W.E., Law, J., 1992. *Shaping Technology/Building Society: Studies in Sociotechnical Change*. Cambridge, MA: MIT Press.

Bijoor, S., 2007. *Thailand Goes Nuclear? Considerations and Costs* [PowerPoint presentation]. 13 August.

Blaikie, P.M., 2012. Should Some Political Ecology Be Useful? The Inaugural Lecture for the Cultural and Political Ecology Specialty Group, Annual Meeting of the Association of American Geographers, April 2010. *Geoforum*, 43 (2), 231–9.

Blaikie, P.M., Brookfield, H.C., 1987. *Land Degradation and Society*. London: Methuen.

Boonpotipukdee, S., 2011. *Thailand Country Report 2010* [PowerPoint presentation]. FNCA Project Leaders Meeting 2010: Public Information of Nuclear Energy, Vietnam.

Bouzarovski, S., 2009a. East-Central Europe's Changing Energy Landscapes: A Place for Geography. *Area*, 41 (4), 452–63.

———, 2009b. *Towards a Critical Geography of the 'Energy Transition'*. ESRC Seminar on Geographies of Energy Transition, University of Leicester, Leicester.

Bradshaw, M.J., 2010. Global Energy Dilemmas: A Geographical Perspective. *The Geographical Journal*, 176 (4), 275–90.

Breukers, S., Wolsink, M., 2007. Wind Energy Policies in the Netherlands: Institutional Capacity-Building for Ecological Modernisation. *Environmental Politics*, 16 (1), 92–112.

Bridge, G., 2011. Past Peak Oil: Political Economy of Energy Crises, in: Peet, R., Robbins, P., Watts, M. (Eds), *Global political ecology*. New York: Routledge, pp. 307–24.

Bridge, G., Bouzarovski, S., Bradshaw, M., Eyre, N., 2013. Geographies of Energy Transition: Space, Place and the Low-Carbon Economy. *Energy Policy*, 53 (February), 331–40.

Brockington, D., 2012. A Radically Conservative Vision? The Challenge of UNEP's Towards a Green Economy. *Development and Change*, 43 (1), 409–22.

Brundtland, G.H., 1987. *Our Common Future*. Oxford: Oxford University Press.

Bryant, R.L., 1998. Power, Knowledge and Political Ecology in the Third World: A Review. *Progress in Physical Geography*, 22 (1), 79–94.

Bryant, R.L., Bailey, S., 1997. *Third World Political Ecology*. London: Routledge.

Bulkeley, H., Broto, V.C., Hodson, M., Marvin, S., 2011. *Cities and Low Carbon Transitions*. London: Routledge.

Buttel, F.H., 2000. Ecological Modernization as Social Theory. *Geoforum*, 31 (1), 57–65.

Buttel, F.H., Spaargaren, G., Mol, A.P.J., 2006. Epilogue: Environmental Flows and Twenty-First-Century Environmental Social Sciences, in: Spaargaren, G., Mol, A.P.J., Buttel, F.H. (Eds), *Governing Environmental Flows: Global Challenges to Social Theory*. Cambridge, MA: MIT Press, pp. 351–77.

Cabraal, R.A., Barnes, D.F., Agarwal, S.G., 2005. Productive Uses of Energy for Rural Development. *Annual Review of Environment and Resources*, 30 (1), 117–44.

Carruthers, B.G., 2005. Historical Sociology and the Economy: Actors, Networks, and Context, in: Adams, J., Clemens, E.S., Orloff, A.S. (Eds), *Remaking Modernity: Politics, History and Sociology*. Durham: Duke University Press, pp. 333–54.

Castells, M., 1996. *The Rise of the Network Society*. New York: Wiley-Blackwell.

Castree, N., 2005. The Epistemology of Particulars: Human Geography, Case Studies and 'Context'. *Geoforum*, 36 (5), 541–4.

———, 2006. From Neoliberalism to Neoliberalisation: Consolations, Confusions, and Necessary Illusions. *Environment and Planning A*, 38 (1), 1–6.

———, 2008a. Neoliberalising Nature: Processes, Effects, and Evaluations. *Environment and planning A*, 40 (1), 153–73.

———, 2008b. Neoliberalising Nature: the Logics of Deregulation and Reregulation. *Environment and planning. A*, 40 (1), 131–52.

———, 2009. Researching Neoliberal Environmental Governance: A Reply to Karen Bakker. *Environment and Planning A*, 41 (8), 1,788–94.

Cavendish, R., 1999. Siam Officially Renamed Thailand. *HistoryToday*, 49 (1).

Chakrabarty, D., 2000. *Provincializing Europe: Postcolonial Thought and Historical Difference*. Princeton: Princeton University Press.

Christoff, P., 1996. Ecological Modernisation, Ecological Modernities. *Environmental Politics*, 5 (3), 476–500.

Coenen, L., Benneworth, P., Truffer, B., 2012. Toward a Spatial Perspective on Sustainability Transitions. *Research Policy*, 41 (6), 968–79.

Coutard, O., Zimmerman, R., Hanley, R., 2005. *Sustaining Urban Networks: The Social Diffusion of Large Technical Systems*. London: Routledge.

Deaton, A., 2005. Measuring Poverty in a Growing World (or Measuring Growth in a Poor World). *Review of Economics and Statistics*, 87 (1), 1–19.

DeLanda, M., 2006. *A New Philosophy of Society: Assemblage Theory and Social Complexity*. London: Continuum.

Delanty, G., 2007. Modernity, in: Ritzer, G. (Ed.), *The Blackwell Encyclopedia of Sociology*. Malden, MA: Blackwell, pp. 3,068–71.

Derrida, J., 1972. Structure, Sign and Play in the Discourse of Human Sciences, in: Macksey, R., Donato, E. (Eds), *The Structuralist Controversy: The Languages of*

Criticism and the Sciences of Man. Baltimore: Johns Hopkins University Press, pp. 223–42.

Dodd, N., 1999. *Social Theory and Modernity*. Cambridge: Polity Press.

Dryzek, J.S., 2005. *The Politics of the Earth: Environmental Discourses*. 2nd edition. Oxford: Oxford University Press.

EdL, 2011a. *About EDL: Historicity Significance*. Vientiane: Electricité du Laos. [online] Available at: http://www.edl.com.la/en/index.php?option=com_content&view=article&id=19&Itemid=28 [Accessed 27 October 2011].

———, 2011b. *Electricité du Laos: Annual Report 2010*. Vientiane: System Planning Office, Technical Department, Electricité du Laos.

———, 2011c. *Electricity Statistics 2010*. Vientiane: Statistic-Planning Office, Electricité du Laos.

———, 2012a. *Development Plan for Electricity Generation until 2012*. Vientiane: Electricité du Laos. [online] Available at: http://www.edl.com.la/file_upload/documents/edl-plan3.jpg [Accessed 30 November 2012].

———, 2012b. *Electricity Statistics 2011*. Vientiane: Statistic-Planning Office, Business-Finance Department, Electricité du Laos.

EGAT, 2010. *Summary of Thailand Power Development Plan 2010–2030*. Bangkok: System Planning Division, EGAT.

———, 2012. *Bhumibol Dam – History of Dams* [เขื่อนภูมิพล ประวัติเขื่อน]. Bangkok: Electricity Generating Authority Thailand (EGAT). [online] Available at: http://www.bhumiboldam.egat.com/historydam.html [Accessed 28 November 2012].

EGCO, 2014. *EGCO – Corporate Profile – History*. Bangkok: Electricity Generating Public Company Limited. [online] Available at: http://www.egco.com/en/corperate_profile_history.asp [Accessed 11 November 2014].

Eisenstadt, S.N., 2000. Multiple Modernities. *Daedalus*, 129 (1), 1–29.

EPPO, 2012. *IPP – SPP – VSPP*. March. Bangkok: Energy Policy and Planning Office (EPPO). [online] Available at: http://www.eppo.go.th/power/data/index.html [Accessed 8 August 2012].

———, 2013. *Energy Statistics*. Bangkok: EPPO. [online] Available at: http://www.eppo.go.th/info/index-statistics.html [Accessed 11 November 2014].

———, 2014a. *Energy statistics – 5. Electricity – Table 5.2–1: Power Generation by Type of Fuel*. Bangkok: EPPO. [online] Available at: http://www.eppo.go.th/info/5electricity_stat.htm [Accessed 11 November 2014].

———, 2014b. *Energy statistics – 5. Electricity – Table 5.5: Import and Export*. Bangkok: EPPO. [online] Available at: http://www.eppo.go.th/info/5electricity_stat.htm [Accessed 11 November 2014].

Escobar, A., 1995. *Encountering Development: The Making and Unmaking of the Third World*. Princeton: Princeton University Press.

———, 1996. Construction Nature: Elements for a Post-structuralist Political Ecology. *Futures*, 28 (4), 325–43.

————, 1999. After Nature: Steps to an Antiessentialist Political Ecology. *Current Anthropology*, 40 (1), 1–30.

————, 2010. Postconstructivist Political Ecologies, in: Redclift, M.R., Woodgate, G. (Eds), *The International Handbook of Environmental Sociology*. Second Edition. Cheltenham: Edward Elgar, pp. 91–105.

Evans, G., 1998. *The Politics of Ritual and Remembrance: Laos since 1975*. Honolulu: University of Hawai'i Press.

————, 2002. *A Short History of Laos: The Land in Between*. Chiang Mai: Silkworm Books.

————, 2012. *A Short History of Laos: The Land in Between*. Revised edition. Chiang Mai: Silkworm Books.

Feeny, D., 1982. *The Political Economy of Productivity: Thai Agricultural Development, 1880–1975*. Vancouver: University of British Columbia Press.

————, 1989. The Decline of Property Rights in Man in Thailand, 1800–1913. *The Journal of Economic History*, 49 (2), 285–96.

Fischer, S., Sahay, R., 1996. Stabilization and Growth in Transition Economies: The Early Experience. *The Journal of Economic Perspectives*, 10 (2), 45–66.

Fischer-Kowalski, M., 2011. Analyzing Sustainability Transitions as a Shift Between Socio-metabolic Regimes. *Environmental Innovation and Societal Transitions*, 1 (1), 152–9.

Fischer-Kowalski, M., Haberl, H., 2007. *Socioecological Transitions and Global Change: Trajectories of Social Metabolism and Land Use*. Cheltenham: Edward Elgar.

Fischer-Kowalski, M., Rotmans, J., 2009. Conceptualizing, Observing and Influencing Social-Ecological Transitions. *Ecology and Society (online)*, 14 (2), 3.

Fisher, D.R., Freudenburg, W.R., 2001. Ecological Modernization and Its Critics: Assessing the Past and Looking Toward the Future. *Society & Natural Resources*, 14 (8), 701–9.

Foran, T., 2006. *Thailand's Politics of Power System Planning and Reform*. Chiang Mai: M-Power and IUCN.

————, 2007. *Rivers of Contention: Pak Mun Dam, Electricity Planning, and State-Society Relations in Thailand, 1932–2004*. Sydney: Department of Geosciences, University of Sydney, p. 377.

Foran, T., du Pont, P., Parinya, P., Phumaraphand, N., 2010. Securing Energy Efficiency as a High Priority: Scenarios for Common Appliance Electricity Consumption in Thailand. *Energy Efficiency*, 3 (4), 347–64.

Foran, T., Manorom, K., 2009. Pak Mun Dam: Perpetually Contested? in: Molle, F., Foran, T., Kakonen, M. (Eds), *Contested Waterscapes in the Mekong Region : Hydropower, Livelihoods and Governance*. London: Earthscan, pp. 55–80.

Forsyth, T., 2001. Critical Realism and Political Ecology, in: Stainer, A., Lopez, G. (Eds), *After Postmodernism: Critical Realism?* London: Athlone, pp. 146–54.

————, 2003. *Critical Political Ecology: The Politics of Environmental Science*. London: Routledge.

———, 2011. Politicizing environmental explanations: what can political ecology learn from sociology and philosophy of science? in: Goldman, M.J., Nadasdy, P., Turner, M.D. (Eds), *Knowing Nature: Conversations At the Intersection of Political Ecology and Science Studies.* Chicago: University of Chicago Press, p. 367.

Frijns, J., Phuong, P.T., Mol, A.P., 2000. Developing Countries: Ecological Modernisation Theory and Industrialising Economies: The Case of Viet Nam. *Environmental politics,* 9 (1), 257–92.

Fullbrook, D., 2007. *Contract Farming in Lao PDR: Cases and Questions.* Vientiane: Laos Extension for Agriculture Project (LEAP).

Gaonkar, D.P., 1999. On Alternative Modernities. *Public Culture,* 11 (1), 1–18.

Gaonkar, D.P. (Ed.), 2001. *Alternative Modernities.* Durham, NC: Duke University Press.

Garrett, B.W., 2004. *Comments on Indicative Master Plan on Power Interconnection in GMS Countries.* Bangkok: Palang Thai.

Geels, F.W., 2002. Technological Transitions as Evolutionary Reconfiguration Processes: A Multi-level Perspective and a Case-Study. *Research Policy,* 31 (8–9), 1,257–74.

———, 2004. From Sectoral Systems of Innovation to Socio-Technical Systems: Insights About Dynamics and Change from Sociology and Institutional Theory. *Research Policy,* 33 (6–7), 897–920.

———, 2005. Processes and Patterns in Transitions and System Innovations: Refining the Co-Evolutionary Multi-Level Perspective. *Technological Forecasting and Social Change,* 72 (6), 681–96.

———, 2007. Feelings of Discontent and the Promise of Middle Range Theory for STS – Examples from Technology Dynamics. *Science Technology & Human Values,* 32 (6), 627–51.

———, 2010. Ontologies, Socio-technical Transitions (to Sustainability), and the Multi-level Perspective. *Research Policy* (4), 495–510.

———, 2011. The Multi-level Perspective on Sustainability Transitions: Responses to Seven Criticisms. *Environmental Innovation and Societal Transitions,* 1 (1), 24–40.

Geels, F.W., Hekkert, M.P., Jacobsson, S., 2008. The Dynamics of Sustainable Innovation Journeys. *Technology Analysis and Strategic Management,* 20 (5), 521–36.

Geels, F.W., Schot, J., 2010. The Dynamics of Transitions: A Socio-technical Perspective, in: Grin, J., Rotmans, J., Schot, J.W. (Eds), *Transitions to Sustainable Development: New Directions in the Study of Long Term Transformative Change.* New York: Routledge, pp. 9–101.

Gibbs, D., 2000. Ecological Modernisation, Regional Economic Development and Regional Development Agencies. *Geoforum,* 31 (1), 9–19.

————, 2006. Prospects for an Environmental Economic Geography: Linking Ecological Modernization and Regulationist Approaches. *Economic Geography*, 82 (2), 193–215.

Gibson-Graham, J.K., 1996. *The End of Capitalism (As We Knew It): A Feminist Critique of Political Economy*. Cambridge, MA: Blackwell.

————, 2008. Diverse Economies: Performative Practices for 'Other Worlds'. *Progress in Human Geography*, 32 (5), 613–32.

Giddens, A., 1984. *The Constitution of Society: Outline of the Theory of Structuration*. Berkeley: University of California Press.

————, 1990. *The Consequences of Modernity*. Stanford: Stanford University Press.

Girling, J.L.S., 1981. *Thailand: Society and Politics*. London: Cornell University Press.

Glassman, J., 2010. *Bounding the Mekong: The Asian Development Bank, China, and Thailand*. Honolulu: University of Hawai'i Press.

GoL, 2012. *Electric Power Plants in Laos – August 2012*. Vientiane: Department of Energy Promotion and Development, Ministry of Energy and Mines. [online] Available at: http://www.poweringprogress.com//download//Electric_Power_Plants_in_Laos_August_2012.pdf [Accessed 29 November 2012].

————, 2014. Power Projects – Operation Projects. Vientiane: Ministry of Energy and Mines, Department of Energy Business. [online] Available at: http://www.poweringprogress.org/new/power-projects/operation [Accessed 12 November 2014].

Goodman, D., Redclift, M.R., 1982. *From Peasant to Proletarian: Capitalist Development and Agrarian Transition*. Oxford: Basil Blackwell.

Graham, S., Marvin, S., 1994. More than Ducts and Wires: Post-Fordism, Cities and Utility Networks, in: Healey, P., Cameron, S., Davoudi, S., Graham, S., Madani-Pour, A. (Eds), *Managing Cities: The New Urban Context*. New York: Chichester.

————, 2001. *Splintering Urbanism: Networked Infrastructures, Technological Mobilities and the Urban Condition*. London: Routledge.

Granovetter, M., 2002 [1985]. Economic Action and Social Structure: The Problem of Embeddedness, in: Biggart, N.W. (Ed.), *Readings in Economic Sociology*. Malden, MA: Blackwell, pp. 63–8.

Greacen, C., 2002. *Small is Pitiful: Community Micro-hydroelectricity and the Politics of Rural Electrification in Thailand* [PhD colloquium PowerPoint presentation]. Berkeley: UC Berkeley.

————, 2004. *The Marginalization of 'Small is Beautiful': Micro-Hydroelectricity, Common Property and the Politics of Rural Electricity Provision in Thailand* [Published PhD Thesis]. Berkeley: Energy and Resources Group, University of California.

Greacen, C., Bijoor, S., 2007. Decentralized Energy in Thailand: An Emerging Light. *World Rivers Review*, 22 (2), 4–5.

Greacen, C., Palettu, A., 2007. Electricity Sector Planning and Hydropower, in: Lebel, L., Dore, J., Daniel, R., Koma, Y.S. (Eds), *Democratizing Water Governance in the Mekong Region*. Chiang Mai: Mekong Press, pp. 93–125.

Greacen, C.S., Greacen, C., 2004. Thailand's Electricity Reforms: Privatization of Benefits and Socialization of Costs and Risks. *Pacific Affairs*, 77 (3), 517–41.

Grin, J., 2010. Understanding Transitions from a Governance Perspective, in: Grin, J., Rotmans, J., Schot, J.W. (Eds), *Transitions to Sustainable Development: New Directions in the Study of Long Term Transformative Change*. New York: Routledge, pp. 221–319.

Grin, J., Rotmans, J., Schot, J.W. (Eds), 2010. *Transitions to Sustainable Development: New Directions in the Study of Long Term Transformative Change*. New York: Routledge.

Hajer, M.A., 1995. *The Politics of Environmental Discourse: Ecological Modernization and the Policy Process*. Oxford: Clarendon Press.

Hansen, U.E., Nygaard, I., 2013. Transnational Linkages and Sustainable Transitions in Emerging Countries: Exploring the Role of Donor Interventions in Niche Development. *Environmental Innovation and Societal Transitions*, 8 (0), 1–19.

———, 2014. Sustainable Energy Transitions in Emerging Economies: The Formation of a Palm Oil Biomass Waste-to-Energy Niche in Malaysia 1990–2011. *Energy Policy*, 66 (0), 666–76.

Hart, G., 2001. Development Critiques in the 1990s: Culs De Sac and Promising Paths. *Progress in Human Geography*, 25 (4), 649–58.

Harvey, D., 1989. The Condition of Postmodernity: An Enquiry into the Origins of Cultural Change. Cambridge: Blackwell.

———, 1996. *Justice, Nature & the Geography of Difference*. Cambridge: Blackwell Publishers.

———, 2005. *A Brief History of Neoliberalism*. New York: Oxford University Press.

Hausman, W.J., Hertner, P., Wilkins, M., 2008. *Global Electrification: Multinational Enterprise and International Finance in the History of Light and Power, 1878–2007*. Cambridge: Cambridge University Press.

Hendriks, C.M., 2008. On Inclusion and Network Governance: The Democratic Disconnect of Dutch Energy Transitions. *Public Administration*, 86 (4), 1,009–31.

———, 2009. Policy Design Without Democracy? Making Democratic Sense of Transition Management. *Policy Sciences*, 42 (4), 341–68.

Herod, A., 2009. *Scale*. Abingdon: Routledge.

Hewison, K., 2006. Thailand: Boom, Bust, and Recovery, in: Rodan, G., Hewison, K., Robison, R. (Eds), *The Political Economy of South-East Asia: Markets, Power and Contestation*. 3rd edition. Melbourne: Oxford University Press, pp. 74–108.

Heynen, N., Kaika, M., Swyngedouw, E. (Eds), 2006. *In the Nature of Cities: Urban Political Ecology and the Politics of Urban Metabolism*. New York: Routledge.

Heynen, N., McCarthy, J., Prudham, S., Robbins, P. (Eds), 2007. *Neoliberal Environments: False Promises and Unnatural Consequences*. London: Routledge.

Hiemstra-van der Horst, G., Hovorka, A.J., 2008. Reassessing the 'energy ladder': Household energy use in Maun, Botswana. *Energy Policy*, 36 (9), 3,333–44.

High, H., 2008. The Implications of Aspirations: Reconsidering Resettlement in Laos. *Critical Asian Studies*, 40 (4), 531–50.

Hillebrand, R., 2013. Climate Protection, Energy Security, and Germany's Policy of Ecological Modernisation. *Environmental Politics*, 22 (4), 664–82.

Hirsch, P., 1988. Dammed Or Damned? Hydropower Versus People's Power. *Critical Asian Studies*, 20 (1), 2–10.

———, 1989. The State in the Village: Interpreting Rural Development in Thailand. *Development and Change*, 20 (1), 35–56.

———, 1996. Large Dams, Restructuring and Regional Integration in Southeast Asia. *Asia Pacific Viewpoint*, 37 1–20.

———, 1998. Dams, Resources and the Politics of Environment in Mainland Southeast Asia, in: Hirsch, P., Warren, C. (Eds), *The politics of Environment in Southeast Asia: Resources and Resistance*. London: Routledge, pp. 55–70.

———, 1999. Dams in the Mekong Region: Scoping Social and Cultural Issues. *Cultural Survival Quarterly*, 23 (3), 37–9.

———, 2007. Civil Society and Interdependencies: Towards a Regional Political Ecology of Mekong Development, in: Connell, J., Waddell, E. (Eds), *Environment, Development and Change in Rural Asia-Pacific: Between Local and Global*. New York: Routledge, pp. 226–46.

———, 2010. The Changing Political Dynamics of Dam Building on the Mekong. *Water Alternatives*, 3 (2), 312–23.

———, 2012. Reviving Agrarian Studies in South-East Asia: Geography on the Ascendancy. *Geographical Research*, 50 (4), 393–403.

Hughes, T.P., 1983. *Networks of Power: Electrification in Western Society, 1880–1930*. Baltimore: Johns Hopkins University Press.

———, 1986. The Seamless Web: Technology, Science, Etcetera, Etcetera. *Social Studies of Science*, 16 (2), 281–92.

ICEM, 2003. *Thailand: National Report on Protected Areas and Development*. Indooroopilly, Queensland, Australia: International Centre for Environmental Management.

IEA, 2000. *World Energy Outlook 2000*. Paris: International Energy Agency.

———, 2012. *World Energy Outlook 2012*. Paris: International Energy Agency.

Ingold, T., 2008. When ANT meets SPIDER: Social Theory for Arthropods, in: Knappett, C., Malafouris, L. (Eds), *Material Agency: Towards a Non-Anthropocentric Approach*. New York: Springer, pp. 209–15.

International Rivers, 2007. *Trading Away the Future: The Mekong Power Grid.* Berkeley: International Rivers.

Jacobs, N., 1971. *Modernization without Development: Thailand as an Asian Case Study.* New York: Praeger.

Jarvis, D.S.L., 2008. *Risk, Regulation & Governance: Institutional Processes and Political Risk in the Thai Energy Sector.* Singapore: Lee Kuan Yew School of Public Policy, National University of Singapore.

Jerndal, R., Rigg, J., 2000. From Buffer State to Crossroads State: Spaces of Human Activity and Integration in the Lao PDR, in: Evans, G. (Ed.), *Laos: Culture and Society.* Singapore: Institute of Southeast Asian Studies, pp. 35–60.

JICA, 1999. *Micro Hydropower Project Portfolio.* Vientiane: JICA, Ministry of Industry and Handicrafts (MIH) and Hydropower Office (HPO).

Kaisti, H., Käkönen, M., 2012. Actors, Interests and Forces Shaping the Energyscape of the Mekong Region. *Forum for Development Studies,* 39 (2), 147–58.

Käkönen, M., Kaisti, H., 2012. The World Bank, Laos and Renewable Energy Revolution in the Making: Challenges in Alleviating Poverty and Mitigating Climate Change. *Forum for Development Studies,* 39 (2), 159–84.

Keil, R., 2003. Urban Political Ecology. *Urban Geography,* 24 (8), 723–38.

Kemp, R., 2010. The Dutch Energy Transition Approach. *International Economics and Economic Policy,* 7 (2), 291–316.

Kemp, R., Loorbach, D., Rotmans, J., 2007a. Transition Management As a Model for Managing Processes of Co-Evolution Towards Sustainable Development. *The International Journal of Sustainable Development & World Ecology,* 14 (1), 78–91.

Kemp, R., Rotmans, J., Loorbach, D., 2007b. Assessing the Dutch Energy Transition Policy: How Does It Deal with Dilemmas of Managing Transitions? *Journal of Environmental Policy and Planning,* 9 (3–4), 315–31.

Kemp, R., Schot, J., Hoogma, R., 1998. Regime Shifts to Sustainability Through Processes of Niche Formation: the Approach of Strategic Niche Management. *Technology Analysis & Strategic Management,* 10 (2), 175–98.

Kern, F., Smith, A., 2008. Restructuring Energy Systems for Sustainability? Energy Transition Policy in the Netherlands. *Energy Policy,* 36 (11), 4,093–103.

Knauft, B.M. (Ed.), 2002. *Critically Modern: Alternatives, Alterities, Anthropologies.* Bloomington: Indiana University Press.

Kolås, Å., 2007. Burma in the Balance: The Geopolitics of Gas*. *Strategic Analysis,* 31 (4), 625–43.

Kuze, N., 2002. *Multi-Organizational Relations in Social Movement: A Case Study of Anti-Power Plant Movements in Hinkrut and Bonok* [in Thai]. Bangkok: Faculty of Political Science, Chulalongkorn University.

———, 2003. *Multi-Organizational Field in Social Movements: A Comparison Between Two Anti-Power Plant Movement Organizations in The Southern Thai Province.*

Occasional Paper No. 11. Tokyo: Centre for Asian Area Studies, Rikkyo University Press.

Labban, M., 2012. Preempting Possibility: Critical Assessment of the IEA's World Energy Outlook 2010. *Development and Change*, 43 (1), 375–93.

Langhelle, O., 2000. Why Ecological Modernization and Sustainable Development Should Not Be Conflated. *Journal of Environmental Policy & Planning*, 2 (4), 303–22.

Lash, S., 1994. Reflexivity and its Doubles: Structure, Aesthetics, Community, in: Beck, U., Giddens, A., Lash, S. (Eds), *Reflexive Modernization: Politics, Tradition and Aesthetics in the Modern Social Order*. Cambridge: Polity Press.

Latour, B., 1987. *Science in Action: How to Follow Scientists and Engineers through Society*. Cambridge, MA: Harvard University Press.

———, 1993. *We Have Never Been Modern*. Cambridge, MA: Harvard University Press.

———, 1999. On Recalling ANT, in: Law, J., Hassard, J. (Eds), *Actor Network Theory and After*. Oxford: Blackwell, pp. 15–25.

———, 2003. Is Re-modernization Occurring-And If So, How to Prove It? *Theory, Culture & Society*, 20 (2), 35–48.

———, 2005. *Reassembling the Social: An Introduction to Actor-Network-Theory*. Oxford: Clarendon.

Law, J., 1992. Notes on the Theory of the Actor-Network: Ordering, Strategy, and Heterogeneity. *Systemic Practice and Action Research*, 5 (4), 379–93.

———, 2004. *After Method: Mess in Social Science Research*. London: Routledge.

———, 2009. Actor Network Theory and Material Semiotics, in: Turner, B.S. (Ed.), *The New Blackwell Companion to Social Theory*. Malden: Wiley-Blackwell, pp. 141–58.

Law, J., Hassard, J. (Eds), 1999. *Actor Network Theory and After*. Oxford: Blackwell.

Lawhon, M., Murphy, J.T., 2011. Socio-Technical Regimes and Sustainability Transitions: Insights from Political Ecology. *Progress in Human Geography*, 36 (3), 354–78.

Lawrence, S., 2009. The Nam Theun 2 Controversy and Its Lessons for Laos, in: Molle, F.O., Foran, T., Kakonen, M. (Eds), *Contested Waterscapes in the Mekong Region: Hydropower, Livelihoods and Governance*. London: Earthscan, pp. 81–114.

Lee, R., 2002. The Demographic Transition: Three Centuries of Fundamental Change. *Journal of Economic Perspectives*, 17 (4), 167–90.

———, 2003. *Tools of Empire or Means of National Salvation? The Railway in the Imagination of Western Empire Builders and Their Enemies in Asia*. York University Institute of Railway Studies & Transport History Working Papers. York: York University Institute of Railway Studies and Transport History.

Li, T.M., 2007a. Practices of Assemblage and Community Forest Management. *Economy and Society*, 36 (2), 263–93.

————, 2007b. *The Will to Improve: Governmentality, Development, and the Practice of Politics*. Durham: Duke University Press.

Loorbach, D., 2010. Transition Management for Sustainable Development: A Prescriptive, Complexity-Based Governance Framework. *Governance*, 23 (1), 161–83.

Lovins, A.B., 2002. *Small is Profitable: The Hidden Economic Benefits of Making Electrical Resources the Right Size*. Snowmass: Rocky Mountain Institute.

Lutzenhiser, L., 1994. Sociology, Energy and Interdisciplinary Environmental Science. *The American Sociologist*, 25 (1), 58–79.

Mann, G., 2009. Should Political Ecology Be Marxist? A Case for Gramsci's Historical Materialism. *Geoforum*, 40 (3), 335–44.

Markard, J., Truffer, B., 2008. Technological Innovation Systems and the Multi-Level Perspective: Towards an Integrated Framework. *Research Policy*, 37 (4), 596–615.

Marsden, T.K., 2009. Sustainability, in: Kitchin, R., Thrift, N. (Eds), *International Encyclopedia of Human Geography*. London: Elsevier, pp. 103–8.

Marston, S.A., Woodward, K., Jones, J.P., 2009. Scale, in: Gregory, D., Johnston, R., Pratt, G., Watts, M.J., Whatmore, S. (Eds), *Dictionary of Human Geography*. Chichester: Blackwell.

Martin, S., Susanto, J., 2014. Supplying Power to Remote Villages in Lao PDR: The Role of Off-Grid Decentralised Energy Options. *Energy for Sustainable Development*, 19 (April), 111–21.

Martinelli, A., 2005. *Global Modernization: Rethinking the Project of Modernity*. London: Sage.

Massey, D., 2005. *For Space*. London: Sage.

Mayntz, R., Hughes, T.P. (Eds), 1988. *The Development of Large Technical Systems*. Frankfurt: Campus Verlag.

Meadowcroft, J., 2009. What about the Politics? Sustainable Development, Transition Management, and Long Term Energy Transitions. *Policy Sciences*, 42 (4), 323–40.

————, 2011. Engaging with the Politics of Sustainability Transitions. *Environmental Innovation and Societal Transitions*, 1 (1), 70–75.

Meadows, D.H., Meadows, D.L., Randers, J., Behrens, W.W, III, 1974. *The Limits to Growth: A Report for the Club of Rome's Project on the Predicament of Mankind*. 2nd edition. New York: New American Library.

Melosi, M., 2006. Energy Transitions in Historical Perspective, in: Dooley, B.M. (Ed.), *Energy and Culture: Perspectives on the Power to Work*. Aldershot: Ashgate, pp. 3–18.

Merme, V., Ahlers, R., Gupta, J., 2014. Private Equity, Public Affair: Hydropower Financing in the Mekong Basin. *Global Environmental Change*, 24 (0), 20–29.

Middleton, C., 2009. *Thailand's Commercial Banks' Role in Financing Dams in Laos and the Case for Sustainable Banking*. Bangkok: International Rivers.

Middleton, C., Garcia, J., Foran, T., 2009. Old and New Hydropower Players in the Mekong Region: Agendas and Strategies, in: Molle, F., Foran, T., Kakonen, M. (Eds), *Contested Waterscapes in the Mekong Region: Hydropower, Livelihoods and Governance*. London: Earthscan, pp. 23–54.

Mills, S., 2003. *Michel Foucault*. London: Routledge.

Ministry of Energy, 2011. *Thailand 20-Year Energy Efficiency Development Plan (2011–2030)*. Bangkok: Ministry of Energy.

———, 2012. *Summary of Thailand Power Development Plan 2012–2030 (PDP2010: Revision 3)*. Bangkok: Energy Policy and Planning Office, Ministry of Energy.

Missingham, B.D., 2003. *The Assembly of the Poor in Thailand: From Local Struggles to National Protest Movement*. Chiang Mai: Silkworm Books.

Mol, A.P.J., 1997. Ecological Modernization: Industrial Transformations and Environmental Reform, in: Woodgate, G., Redclift, M. (Eds), *The International Handbook of Environmental Sociology*. Cheltenham: Edward Elgar.

———, 2001. *Globalization and Environmental Reform: the Ecological Modernization of the Global Economy*. Cambridge, MA: MIT Press.

———, 2010a. Ecological Modernization as a Social Theory of Environmental Reform, in: Redclift, M.R., Woodgate, G. (Eds), *The International Handbook of Environmental Sociology*. Second edition. Cheltenham: Edward Elgar, pp. 63–76.

———, 2010b. Social Theories of Environmental Reform: Towards a Third Generation, in: Gross, M., Heinrichs, H. (Eds), *Environmental Sociology: European Perspectives and Interdisciplinary Challenges*. Dordrecht: Springer, pp. 19–38.

Mol, A.P.J., Sonnenfeld, D.A., 2000. *Ecological Modernisation around the World: Perspectives and Critical Debates*. London: Routledge.

Mol, A.P.J., Sonnenfeld, D.A., Spaargaren, G. (Eds), 2009. *The Ecological Modernisation Reader: Environmental Reform in Theory and Practice*. Abingdon: Routledge.

Mol, A.P.J., Spaargaren, G., Sonnenfeld, D.A., 2014. Ecological Modernization Theory: Taking Stock, Moving Forward, in: Lockie, S., Sonnenfeld, D.A., Fisher, D.R. (Eds), *International Handbook of Social and Environmental Change*. London: Routledge, pp. 15–30.

Molle, F., Foran, T., Floch, P., 2009a. Introduction: Changing Waterscapes in the Mekong Region – Historical Background and Context, in: Molle, F., Foran, T., Kakonen, M. (Eds), *Contested Waterscapes in the Mekong Region: Hydropower, Livelihoods and Governance*. London: Earthscan, p. 426.

Molle, F., Foran, T., Kakonen, M. (Eds), 2009b. *Contested Waterscapes in the Mekong Region: Hydropower, Livelihoods and Governance*. London: Earthscan.

Nartsupha, C., Prasartset, S., Chenvidyakarn, M. (Eds), 1978. *The Political Economy of Siam, 1910–1932*. 2nd edition. Bangkok: Social Science Association of Thailand.

Neumann, R.P., 2005. *Making Political Ecology*. London: Hodder Arnold.

————, 2009. Political Ecology: Theorizing Scale. *Progress in Human Geography*, 33 (3), 398–406.

Nicolas, F., 2009. *ASEAN Energy Cooperation: An Increasingly Daunting Challenge.* Paris: Institut Francais des Relations Internationales.

NSO, 2012. *Rate of population 6 years of age and over by viewing television, by age group: 1989, 1994, 2003 and 2008.* Accessed 21 March 2012. Bangkok: National Statistical Office, Ministry of Information and Communication Technology.

O'Connor, P.A., 2010. Energy Transitions. *The Pardee Papers No. 12.* Boston: The Frederick S. Pardee Center for the Study of the Longer-Range Future, Boston University.

Oehlers, A., 2006. A critique of ADB policies towards the Greater Mekong Sub-region. *Journal of Contemporary Asia*, 36 (4), 464–78.

Oxford English Dictionary, 2002a. Modernity. [online] Available at: http://www.oed.com/view/Entry/120626?redirectedFrom=modernity#eid. *Oxford English Dictionary.* 3rd edition. Oxford: Oxford University Press.

————, 2002b. Transition, n. [online] Available at: http://www.oed.com/view/Entry/204815. *Oxford English Dictionary.* 3rd edition. Oxford: Oxford University Press.

————, 2012a. Capitalism, n.2. [online] Available at: http://www.oed.com/view/Entry/27454. *Oxford English Dictionary.* 3rd edition. Oxford: Oxford University Press..

————, 2012b. Sustainability (b. spec.). [online] Available at: http://www.oed.com/view/Entry/299890. *Oxford English Dictionary.* 3rd edition. Oxford: Oxford University Press.

Paling, W., Winter, T., 2011. Sustainability, Consumption and the Household in Developing World Contexts, in: Lane, R., Gorman-Murray, A. (Eds), *Material Geographies of Household Sustainability.* Farnham: Ashgate, pp. 51–67.

Palmer, D.D., 1997. *Structuralism and Poststructuralism for Beginners.* London: Writers and Readers.

Pasqualetti, M.J., 2011. The Geography of Energy and the Wealth of the World. *Annals of the Association of American Geographers*, 101 (4), 971–80.

Peet, R., Robbins, P., Watts, M. (Eds), 2011. *Global Political Ecology.* New York: Routledge.

Peet, R., Watts, M. (Eds), 2004. *Liberation Ecologies: Environment, Development, Social Movements.* 2nd edition. London: Routledge.

Personal Communication, 2012. Personal Communication Helvetas, 13 September.

Phetsiriseng, I., 2001. *Preliminary Assessment of Illegal Labour Migration and Trafficking in Children and Women for Labour Exploitation.* Vientiane: International Labour Organization.

Pholsena, V., 2006. *Post-War Laos: The Politics of Culture, History, and Identity.* New York: Cornell University Press.

Pholsena, V., Banomyong, R., 2006. *Laos: From Buffer State to Crossroads?* Chiang Mai: Mekong Press and Silkworm Books.

Phomsoupha, X., 2009. Hydropower Development Plans and Progress in Lao PDR. *Hydro Nepal*, 4 (January), 15–17.

Phongpaichit, P., 1980. The Open Economy and Its Friends: The 'Development' of Thailand. *Pacific Affairs*, 53 (3), 440–60.

Phraxayavong, V., 2009. *History of Aid to Laos: Motivations and Impacts*. Chiang Mai: Mekong Press.

Prachuab Provincial Office, 2011. *Summary Report Prachuab Khiri Khan Province 2011* [บรรยายสรุป จังหวัดประจวบคีรีขันธ์ประจำปี 2554]. Prachuab Khiri Khan: Information technology, and Communication Division, Prachuab Khiri Khan provincial office [กลุ่มงานข้อมูลสารสนเทศและการสื่อสาร สำนักงานจังหวัดประจวบคีรีขันธ์].

Pred, A.R., Watts, M., 1992. *Reworking Modernity: Capitalisms and Symbolic Discontent*. New Brunswick: Rutgers University Press.

Promjittiphong, C., 2005. *Foreigner Tourist Satisfaction on Homestay Services in Royal Project Development Center Teen – Tok, Chiang Mai Province* [MSc thesis]. Bangkok: Faculty of Environmental and Resource Studies, Mahidol University.

Prudham, S., 2009. Sustainability, in: Gregory, D., Johnston, R., Pratt, G., Watts, M.J., Whatmore, S. (Eds), *Dictionary of Human Geography*. Malden: Blackwell, pp. 737–8.

Raven, R., 2007. Niche Accumulation and Hybridisation Strategies in Transition Processes towards a Sustainable Energy System: An Assessment of Differences and Pitfalls. *Energy Policy*, 35 (4), 2,390–400.

Raven, R., Schot, J.W., Berkhout, F., 2012. Space and Scale in Socio-Technical Transitions. *Environmental Innovation and Societal Transitions*, 4 (September), 63–78.

Reckwitz, A., 2002. Toward a Theory of Social Practices. *European Journal of Social Theory*, 5 (2), 243.

Redclift, M.R., 2005. Sustainable Development (1987–2005): An Oxymoron Comes of Age. *Sustainable Development*, 13 (4), 212–27.

Rigg, J., 2001. *More than the Soil: Rural Change in Southeast Asia*. Harlow: Prentice Hall.

———, 2003. *Southeast Asia: the Human Landscape of Modernization and Development*. 2nd edition. London: Routledge.

———, 2005. *Living with Transition in Laos: Market Integration in Southeast Asia*. London: Routledge.

———, 2007. *An Everyday Geography of the Global South*. London: Routledge.

Rigg, J., Jerndal, R., 1996. Plenty in the Context of Scarcity: Forest Management in Laos, in: Parnwell, M.J., Bryant, R.L. (Eds), *Environmental Change in South-East Asia: People, Politics and Sustainable Development*. London: Routledge.

Rigg, J., Salamanca, A., Parnwell, M., 2012. Joining the Dots of Agrarian Change in Asia: A 25 Year View from Thailand. *World Development*, 40 (7), 1,469–81.

Rip, A., Kemp, R., 1998. Technological Change, in: Rayner, S., Malone, E. (Eds), *Human Choice and Climate Change*. Columbus: Battelle Press, pp. 327–92.

RISE, 2008. *Report of Observation the village grid in Ban Nam Kha*. Phonesavan: Rural Income through Sustainable Energy (RISE) Project.

————, 2009. *Participatory Baseline Survey in Ban Nam Kha, Phaxay District, Xieng Khuang Province: Result Report – January 2009*. Phonesavan: Rural Income through Sustainable Energy (RISE) Project.

Robbins, P., 2012. *Political Ecology: A Critical Introduction*. 2nd edition. Malden, MA: Blackwell.

Rostow, W.W., 1960. *The Stages of Economic Growth: A Non-Communist Manifesto*. Cambridge: Cambridge University Press.

Rotmans, J., Kemp, R., 2008. Detour Ahead: A Response to Shove and Walker About the Perilous Road of Transition Management. *Environment and Planning A: International Journal of Urban and Regional Research*, 40 (4), 1,006–12.

Roy, A., 1999. *The Cost of Living: The Greater Common Good and the End of Imagination*. London: Flamingo.

Ryder, G., 2003. *Behind the ASEAN Power Grid: Analysis of the Asian Development Bank's Master Plan for Regional Power Interconnections and Power Trade in the Greater Mekong Subregion*. Toronto: Probe International.

————, 2004. *Ten Reasons Why the World Bank Should Not Finance the Nam Theun 2 Power Company in Lao PDR*. Toronto: Probe International Backgrounder, Probe International.

Sahakian, M.D., 2010. Understanding Household Energy Consumption Patterns: When 'West is Best' in Metro Manila. *Energy Policy*, 39 (2), 596–602.

————, 2014. *Keeping Cool in Southeast Asia: Energy Consumption and Urban Air-Conditioning*. New York: Palgrave.

Sahakian, M.D., Steinberger, J.K., 2011. Energy Reduction through a Deeper Understanding of Household Consumption. *Journal of Industrial Ecology*, 15 (1), 31–48.

Savada, A.M. (Ed.), 1994. *Laos: A Country Study*. Washington: GPO for the Library of Congress. [online] Available at: http://countrystudies.us/laos/78.htm.

Sayer, R.A., 1997. Essentialism, Social Constructionism, and Beyond. *Sociological Review*, 45 (3), 453–87.

————, 2000a. For Postdisciplinary Studies: Sociology and the Curse of Disciplinary Parochialism/Imperialism, in: Eldridge, J., MacInnes, J., Scott, S., Warhurst, C., Witz, A. (Eds), *For Sociology: Legacies and Prospects*. Durham: Sociologypress, pp. 83–92.

————, 2000b. *Realism and Social Science*. London: Sage.

Schnaiberg, A., 1980. *The Environment: From Surplus to Scarcity*. Oxford: Oxford University Press.

————, 1997. The Political Economy of Environmental Problems and Policies: Consciousness, Conflict, and Control Capacity, in: Freese, L. (Ed.), *Advances in Human Ecology, volume III*. Greenwich, CT: JAI Press.

Schot, J., Geels, F.W., 2008. Strategic Niche Management and Sustainable Innovation Journeys: Theory, Findings, Research Agenda, and Policy. *Technology Analysis & Strategic Management*, 20 (5), 537–54.

Schumacher, E.F., 1975. *Small is Beautiful: Economics as if People Mattered*. New York: Harper & Row.

Scott, J.C., 2009. *The Art of Not Being Governed*. New Haven: Yale University Press.

Scrase, I., Smith, A., 2009. The (Non-)Politics of Managing Low Carbon Socio-Technical Transitions. *Environmental Politics*, 18 (5), 707–26.

Sheppard, E., Leitner, H., 2010. Quo Vadis Neoliberalism? The Remaking of Global Capitalist Governance After the Washington Consensus. *Geoforum*, 41 (2), 185–94.

Sheppard, E., McMaster, R.B., 2004. *Scale and Geographic Inquiry: Nature, Society, and Method*. Malden, MA: Blackwell.

Shoemaker, B., 1998. *Trouble on the Theun-Hinboun: A Field Report On the Socio-Economic and Environmental Effects of the Nam Theun-Hinboun Hydropower Project in Laos*. Toronto: Probe International.

Shove, E., 1997. Revealing the Invisible: Sociology, Energy and the Environment, in: Woodgate, G., Redclift, M. (Eds), *The International Handbook of Environmental Sociology*. Cheltenham: Edward Elgar, pp. 261–73.

————, 2003a. *Comfort, Cleanliness and Convenience: The Social Organization of Normality*. Oxford: Berg.

————, 2003b. Converging Conventions of Comfort, Cleanliness and Convenience. *Journal of Consumer Policy*, 26 (4), 395–418.

————, 2004a. Efficiency and Consumption: Technology and Practice. *Energy & Environment*, 15 (6), 1,053–65.

————, 2004b. Sustainability, System Innovation and the Laundry, in: Elzen, B., Geels, F.W., Green, K. (Eds), *System Innovation and the Transition to Sustainability: Theory, Evidence and Policy*. Cheltenham: Edward Elgar, pp. 76–94.

Shove, E., Lutzenhiser, L., Guy, S., Hackett, B., Wilhite, H., 1998. Energy and Social Systems, in: Rayner, S., Malone, E. (Eds), *Human Choice and Climate Change*. Columbus: Batelle Press, pp. 291–325.

Shove, E., Pantzar, M., 2005. Consumers, Producers and Practices Understanding the invention and reinvention of Nordic walking. *Journal of consumer culture*, 5 (1), 43–64.

Shove, E., Pantzar, M., Watson, M., 2012. *The Dynamics of Social Practice: Everyday Life and How It Changes*. London: Sage.

Shove, E., Walker, G., 2007. CAUTION! Transitions Ahead: Politics, Practice and Transition Management. *Environment and Planning A*, 39 (4), 763–70.

————, 2008. Transition Management™ and the Politics of Shape Shifting. *Environment and Planning A: International Journal of Urban and Regional Research*, 40 (4), 1,012–14.

————, 2010. Governing Transitions in the Sustainability of Everyday Life. *Research Policy*, 39 (4), 471–6.

————, 2014. What is Energy for? Social Practice and Energy Demand. *Theory, Culture & Society*, 31 (5), 41–58.

Shove, E., Walker, G., Brown, S., 2013. Transnational Transitions: The Diffusion and Integration of Mechanical Cooling. *Urban Studies*, 51 (7), 1,506–19.

Shrestha, R., Kumar, S., Sharma, S., Todoc, M., 2004. Institutional Reforms and Electricity Access: Lessons from Bangladesh and Thailand. *Energy for Sustainable Development*, 8 (4), 41–53.

Simpson, A., 2007. The Environment – Energy Security Nexus: Critical Analysis of An Energy 'Love Triangle' in Southeast Asia. *Third World Quarterly*, 28 (3), 539–54.

Singh, S., 2009. World Bank-Directed Development? Negotiating Participation in the Nam Theun 2 Hydropower Project in Laos. *Development and Change*, 40 (3), 487–507.

Smil, V., 1994. *Energy in World History*. Boulder: Westview Press.

Smith, A., Kern, F., 2009. The Transitions Storyline in Dutch Environmental Policy. *Environmental Politics*, 18 (1), 78–98.

Smith, A., Stirling, A., 2010. The Politics of Social-Ecological Resilience and Sustainable Socio-technical Transitions. *Ecology and Society*, 15 (1), 11.

Smits, M., Bush, S.R., 2010. A Light Left in the Dark: The Practice and Politics of Pico-Hydropower in the Lao PDR. *Energy Policy*, 38 (1), 116–27.

Smits, M., Middleton, C., 2014. New Arenas of Engagement at the Water Governance-Climate Finance Nexus? An Analysis of Boom and Bust of Hydropower CDM in Vietnam. *Water Alternatives*, 7 (3), 561–83.

Sneddon, C., Fox, C., 2006. Rethinking Transboundary Waters: A Critical Hydropolitics of the Mekong Basin. *Political Geography*, 25 (2), 181–202.

Solomon, B.D., Pasqualetti, M.J., Luchsinger, D.A., 2003. Energy Geography, in: Gaile, G.L., Willmott, C.J. (Eds), *Geography in America at the Dawn of the 21st Century*. Oxford: Oxford University Press, pp. 302–13.

Sonnenfeld, D.A., 2000. Developing Countries: Contradictions of Ecological Modernisation: Pulp and Paper Manufacturing in South East Asia. *Environmental Politics*, 9 (1), 235–56.

Sovacool, B.K., 2009a. Energy Policy and Cooperation in Southeast Asia: The History, Challenges, and Implications of the Trans-ASEAN Gas Pipeline (TAGP) Network. *Energy Policy*, 37 (6), 2,356–67.

————, 2009b. Reassessing Energy Security and the Trans-ASEAN Natural Gas Pipeline Network in Southeast Asia. *Pacific Affairs*, 82 (3), 467–86.

————, 2012. The Political Economy of Energy Poverty: A Review of Key Challenges. *Energy for Sustainable Development*, 16 (3), 272–82.

Spaargaren, G., 2011. Theories of Practices: Agency, Technology, and Culture: Exploring the Relevance of Practice Theories for the Governance of Sustainable Consumption Practices in the New World-Order. *Global Environmental Change*, 21 (3), 813–22.

Spaargaren, G., Mol, A.P.J., 1992. Sociology, Environment, and Modernity: Ecological Modernization As a Theory of Social Change. *Society & Natural Resources*, 5 (4), 323–44.

Spaargaren, G., Mol, A.P.J., Bruyninckx, H., 2006a. Introduction: Governing Environmental Flows in Global Modernity, in: Spaargaren, G., Mol, A.P.J., Buttel, F.H. (Eds), *Governing Environmental Flows: Global Challenges to Social Theory*. Cambridge, MA: MIT Press, pp. 351–77.

Spaargaren, G., Mol, A.P.J., Buttel, F.H. (Eds), 2006b. *Governing Environmental Flows: Global Challenges to Social Theory*. Cambridge: MIT Press.

Späth, P., Rohracher, H., 2014. Beyond Localism: The Spatial Scale and Scaling in Energy Transitions, in: Padt, F., Opdam, P., Polman, N., Termeer, C. (Eds), *Scale-Sensitive Governance of the Environment*. Chichester: Wiley Blackwell, pp. 106–21.

Star, S.L., 1999. The Ethnography of Infrastructure. *American Behavioral Scientist*, 43 (3), 377–91.

Starostina, N., 2010. Ambiguous Modernity: Representations of French Colonial Railways in the Third Republic, in: Neulander, J., Walz, R. (Eds), *Annual Meeting of the Western Society for French History*. Ann Arbor: University of Michigan Library.

Stifel, L.D., 1976. Technocrats and Modernization in Thailand. *Asian Survey*, 16 (12), 1,184–96.

Stiglitz, J.E., 2003. *Globalization and Its Discontents*. New York: W.W. Norton.

Stott, P.A., Sullivan, S. (Eds), 2000. *Political Ecology: Science, Myth and Power*. London: Arnold.

Strohmayer, U., 2009. Modernity, in: Gregory, D., Johnston, R., Pratt, G., Watts, M.J., Whatmore, S. (Eds), *Dictionary of Human Geography*. Malden, MA: Blackwell, pp. 471–4.

Stuart-Fox, M., 1997. *A History of Laos*. Cambridge: Cambridge University Press.

Sukkumnoed, D., 2007. *Better Power for Health: Healthy Public Policy and Sustainable Energy in the Thai Power Sector* [Unpublished PhD Thesis]. Aalborg: Department of Development and Planning, Aalborg University.

Summerton, J. (Ed.), 1994. *Changing Large Technical Systems*. Boulder: Westview Press.

Susanto, J., Smits, M., 2010. Towards a Locally Adapted Rural Electrification Assessment Framework: A Case Study of the Lao PDR. *GMSTEC*

2010: *International Conference for a Sustainable Greater Mekong Subregion.* Bangkok, Thailand.

Swyngedouw, E., 1997. Neither Global Nor Local: 'Glocalization'and the Politics of Scale, in: Cox, K.R. (Ed.), *Spaces of Globalization: Reasserting the Power of the Local.* New York: Guilford Press, pp. 137–66.

———, 2000. The Marxian Alternative: Historical-Geographical Materialism and the Political Economy of Capitalism, in: Sheppard, E., Barnes, T.J. (Eds), *A Companion to Economic Geography.* Malden, MA: Blackwell, pp. 41–59.

———, 2010. Apocalypse Forever? Post-political Populism and the Spectre of Climate Change. *Theory, Culture & Society*, 27 (2–3), 213–32.

Swyngedouw, E., Heynen, N.C., 2003. Urban Political Ecology, Justice and the Politics of Scale. *Antipode*, 35 (5), 898–918.

Ten Brummelhuis, H., 2005. *King of the Waters: Homan van der Heide and the Origin of Modern Irrigation in Siam.* Leiden: KITLV Press.

The Economist, 1993. Laos Joins the World: Laos Plots Its Path to Riches. *The Economist*, 6 November. London: The Economist Newspaper Ltd.

The Wall Street Journal, 2011. Laos Looks to Become a New Market Frontier. *The Wall Street Journal*, 11 January. Dow Jones.

Thongplew, N., Spaargaren, G., Van Koppen, C.S.A., 2013. Greening Consumption at the Retail Outlet: the Case of the Thai Appliance Industry. *International Journal of Sustainable Development & World Ecology*, 21 (2), 99–110.

Thongplew, N., van Koppen, C.S.A., Spaargaren, G., 2014. Companies Contributing to the Greening of Consumption: Findings from the Dairy and Appliance Industries in Thailand. *Journal of Cleaner Production*, 75 (0), 96–105.

Thrift, N., 2005. *Knowing Capitalism.* London: Sage.

Toke, D., 2011. *Ecological Modernisation and Renewable Energy.* Basingstoke: Palgrave Macmillan.

Tsing, A.L., 2005. *Friction: An Ethnography of Global Connection.* Princeton: Princeton University Press.

UN, 2010. *Energy for a Sustainable Future: The Secretary-General's Advisory Group on Energy and Climate Change – Summary Report and Recommendations.* New York: United Nations.

UNCTAD, 2011. *The Least Developed Countries Report 2011: The Potential Role of South-South Cooperation for Inclusive and Sustainable Development.* New York: United National Conference on Trade and Development.

UNDP, 2012. *The Lao Development Journey Toward Graduation from LDC Status.* Vientiane: UNDP and Government of Laos.

Urry, J., 2000. *Sociology Beyond Societies: Mobilities for the Twenty-First Century.* London: Routledge.

———, 2003. *Global Complexity.* Cambridge: Polity Press.

———, 2011. *Climate Change and Society.* Cambridge: Polity Press.

Van Vliet, B., 2002. *Greening the Grid: the Ecological Modernisation of Network-Bound Systems* [Unpublished PhD Thesis]. Wageningen: Environmental Policy Group, University of Wageningen.

Vandergeest, P., 2003. Land to Some Tillers: Development-Induced Displacement in Laos. *International Social Science Journal*, 55 (175), 47–56.

Vandergeest, P., Peluso, N.L., 1995. Territorialization and State Power in Thailand. *Theory and Society*, 24 (3), 385–426.

Verbong, G., Geels, F.W., 2007. The Ongoing Energy Transition: Lessons from a Socio-technical, Multi-level Analysis of the Dutch Electricity System (1960–2004). *Energy Policy*, 35 (2), 1,025–37.

Verbong, G., Geels, F.W., Raven, R., 2008. Multi-Niche Analysis of Dynamics and Policies in Dutch Renewable Energy Innovation Journeys (1970–2006): Hype-Cycles, Closed Networks and Technology-Focused Learning. *Technology Analysis & Strategic Management*, 20 (5), 555–73.

Voß, J.-P., Kemp, R., 2006. Sustainability and Reflexive Governance: Introduction, in: Voß, J.-P., Bauknecht, D., Kemp, R. (Eds), *Reflexive Governance for Sustainable Development*. Cheltenham: Edward Elgar, pp. 3–28.

Wagner, P., 2001. *Theorizing Modernity: Inescapability and Attainability in Social Theory*. London: SAGE.

Walker, G., Shove, E., 2007. Ambivalence, Sustainability and the Governance of Socio-Technical Transitions. *Journal of Environmental Policy and Planning*, 9 (3–4), 213–25.

Wallerstein, I., 1974. *The Modern World-System I: Capitalist Agriculture and the Origins of the European World-Economy in the Sixteenth Century*. New York: Academic Press.

Warde, A., Shove, E., Southerton, D., 1998. *Convenience, Schedules and Sustainability*. ESF workshop on sustainable consumption. Lancaster.

Warner, R., 2010. Ecological Modernisation Theory: Towards a Critical Ecopolitics of Change? *Environmental Politics*, 19 (4), 538–56.

Watson, M., 2012. How Theories of Practice Can Inform Transition to a Decarbonised Transport System. *Journal of Transport Geography*, 24 (September), 488–96.

Wattana, S., Sharma, D., Vaiyavuth, R., 2007. Electricity Industry Reforms in Thailand: A Historical Review. *GMSARN International Conference on Sustainable Development: Challenges and Opportunities for GMS*.Bangkok.

Watts, M.J., 1983. *Silent Violence: Food, Famine, & Peasantry in Northern Nigeria*. Berkeley: University of California Press.

———, 2000. Political Ecology, in: Sheppard, E., Barnes, T.J. (Eds), *A Companion to Economic Geography*. Malden, MA: Blackwell, pp. 257–74.

———, 2003. Alternative Modern: Development as Cultural Geography, in: Anderson, K., Domosh, M., Pile, S., Thrift, N. (Eds), *Handbook of Cultural Geography*. London: Sage, pp. 433–53.

————, 2009a. Capitalism, in: Gregory, D., Johnston, R., Pratt, G., Watts, M.J., Whatmore, S. (Eds), *Dictionary of Human Geography*. Malden, MA: Blackwell, pp. 59–64.

————, 2009b. Political ecology, in: Gregory, D., Johnston, R., Pratt, G., Watts, M.J., Whatmore, S. (Eds), *Dictionary of Human Geography*. Malden, MA: Blackwell.

WCD, 2000. *Dams and Development: A New Framework for Decision-Making, Report by the World Commission on Dams*. London: Earthscan.

Wilhite, H., 2008. *Consumption and the Transformation of Everyday Life: A View from South India*. Basingstoke: Palgrave Macmillan.

————, 2012. A Socio-cultural Analysis of Changing Household Electricity Consumption in India, in: Spreng, D., Flüeler, T., Goldblatt, D.L., Minsch, J. (Eds), *Tackling Long-Term Global Energy Problems: The Contributions of Social Science*. Dordrecht: Springer, pp. 97–113.

Wilk, R., 2002. Consumption, Human Needs, and Global Environmental Change. *Global Environmental Change*, 12 (1), 5–13.

Williams, J.H., Dubash, N.K., 2004. Asian Electricity Reform in Historical Perspective. *Pacific Affairs*, 77 (3), 411–36.

Willis, K., 2005. *Theories and Practices of Development*. London: Routledge.

Winichakul, T., 1994. *Siam Mapped: A History of the Geo-body of a Nation*. Honolulu: University of Hawai'i Press.

Wisuttisak, P., 2010. Regulatory Framework of Thai Electricity Sector. *Third Annual Conference on Competition and Regulation in Network Industries*. Brussels, Belgium.

Wolfe, C., 1998. *Critical Environments: Postmodern Theory and the Pragmatics of the 'Outside'*. Minneapolis: University of Minnesota Press.

Wongsawat, Y., Bhuntuvech, C., 2009. Chiang Mai's Experience with Developing Sustainable Tourism. *3rd International Colloquium on Tourism and Leisure*. Bangkok, Thailand.

Woodward, K., Dixon, D., Jones, J.P., III, 2009. Poststructuralism/ Poststructuralist Geographies, in: Kitchin, R., Thrift, N. (Eds), *International Encyclopedia of Human Geography*. London: Elsevier.

World Bank, 1959. *A Public Development Program for Thailand: Report of a Mission Organized by the International Bank for Reconstruction and Development at the Request of the Government of Thailand*. Baltimore: Johns Hopkins Press.

————, 1981a. *Report and Recommendation of the President of the International Development Association to the Executive Directors on a Proposed Development Credit to the Lao People's Democratic Republic for a Nam Ngum Hydroelectric Project*. Manila: World Bank.

————, 1981b. *Staff Appraisal Report Lao People's Democratic Republic Nam Ngum Hydroelectric Project*. Manila: Projects Department, East Asia and Pacific Regional Office, World Bank.

———, 1995. *Implementation Completion Report Lao People's Democratic Republic Southern Provinces Electrification Project*. Washington, DC: Infrastructure Operations Division, Country Department, East Asia and Pacific Region.

———, 2007. *Strategy Note on World Bank Regional Support for the Greater Mekong Sub-Region*. Bangkok: Southeast Asia Country Management Unit, East Asia and Pacific Region, World Bank.

———, 2009. R*ural Electrification Phase II Project in support of the Rural Electrification (APL) Program*. Washington, DC: Infrastructure Unit, Sustainable Development Department, World Bank.

World Bank and AusAID, 2011. *One Goal, Two Paths: Achieving Universal Access to Modern Energy in East Asia and the Pacific*. Washington: World Bank and AusAID.

Wyatt, A.B., 2004. *Infrastructure development and BOOT in Laos and Vietnam: A Case Study of Collective Action and Risk in Transitional Developing Economies* [Unpublished PhD Thesis]. Sydney: School of Geosciences, The University of Sydney.

Wyatt, D.K., 1984. *Thailand: A Short History*. New Haven: Yale University Press.

York, R., Rosa, E.A., Dietz, T., 2010. Ecological Modernization Theory: Theoretical and Empirical Challenges, in: Redclift, M.R., Woodgate, G. (Eds), *The International Handbook of Environmental Sociology*. 2nd edition. Cheltenham: Edward Elgar.

Yu, X., 2003. Regional Cooperation and Energy Development in the Greater Mekong Sub-Region. *Energy Policy*, 31 (12), 1,221–34.

Yuan, Y., Raubal, M., Liu, Y., 2012. Correlating Mobile Phone Usage and Travel Behavior: A Case Study of Harbin, China. *Computers, Environment and Urban Systems*, 36 (2), 118–30.

Zimmerer, K.S., 2011. New Geographies of Energy: Introduction to the Special Issue. *Annals of the Association of American Geographers*, 101 (4), 705–11.

Zimmerer, K.S., Bassett, T. (Eds), 2003. *Political Ecology: An Integrative Approach to Geography and Environment-Development Studies*. New York: Guilford Press.

Index